Planning on the Edge

The rural-urban fringe has been called 'planning's last frontier', an area that has slipped through the gap between town and country, and is in urgent need of attention from policy makers. It is often seen as a degraded area, fit only for locating sewage works, essential service functions and other, necessary but disliked, land uses. But, the fringe is also a dynamic area where a range of urban uses, including large shopping centres, office buildings and business parks and leisure facilities collide with rural uses, such as farming and forestry, outdoor recreation, and conservation.

In *Planning on the Edge* the authors investigate what the rural-urban fringe is and what has underpinned its development. They also examine the context for planning and other forms of policy intervention at the fringe, and how we might, in the future, more effectively manage and plan the edges of towns and cities. *Planning on the Edge* fills an important gap in land use and spatial planning literature, and examines in detail the context for planning in the 'no-man's land' of the rural-urban fringe. Gallent, Andersson and Bianconi are particularly concerned with the legitimate role of planning in a largely unplanned landscape; does the fringe need to be subjected to conventional planning and, therefore, a process of 'sanitization', or might a gentler approach to steering development at the fringe be preferable?

This book challenges the wisdom of over-designing landscapes, arguing that the fringe is an integral and inevitable part of the urban system and a product of largely organic processes. It investigates the way in which landscapes are made through urban containment, alongside the departure from a purely land-use planning model, and the future role of spatial planning at the edge. *Planning on the Edge* will be of interest not only to students of various disciplines, including geography and planning, and researchers, but also to policy makers and planning practitioners.

Planning on the Edge

The context for planning at the rural-urban fringe

Nick Gallent, Johan Andersson and Marco Bianconi

LONDON AND NEW YORK

First published 2006 by Routledge
2 Park Square, Milton Park, Abingdon, Oxon OX14 4RN

Simultaneously published in the USA and Canada
by Routledge
270 Madison Avenue, New York, NY 10016

Routledge is an imprint of the Taylor & Francis Group, an informa business

© 2006 Nick Gallent, Johan Andersson and Marco Bianconi

Typeset in Akzidenz Grotesk by
Keystroke, Jacaranda Lodge, Wolverhampton
Printed and bound in Great Britain by
TJ International, Padstow, Cornwall

All rights reserved. No part of this book may be reprinted or reproduced or utilized in any form or by any electronic, mechanical, or other means, now known or hereafter invented, including photocopying and recording, or in any information storage or retrieval system, without permission in writing from the publishers.

British Library Cataloguing in Publication Data
A catalogue record for this book is available from the British Library

Library of Congress Cataloging in Publication Data
Gallent, Nick.
Planning on the edge : the context for planning at the rural-urban fringe / Nick Gallent, Johan Andersson and Marco Bianconi.
p. cm.
Includes bibliographical references and index.
ISBN 0–415–37571–1 (hb : alk. paper) – ISBN 0–415–40290–5 (pb : alk. paper)
1. City planning–England. 2. Rural-urban relations–England. 3. Mixed-use developments–England. 4. Land use, Rural–England. 5. Land use, Urban–England. I. Title: Rural-urban fringe. II. Andersson, Johan. III. Bianconi, Marco. IV. Title.
HT169.G72E5836 2006
307.1'2160942–dc22

ISBN10: 0–415–37571–1   ISBN13: 978–0–415–37571–9 (hbk)
ISBN10: 0–415–40290–5   ISBN13: 978–0–415–40290–3 (pbk)
ISBN10: 0–203–09919–2   ISBN13: 978–0–203–09919–3 (ebk)

Contents

List of figures	ix
List of tables	xi
Glossary of acronyms	xiii
Preface	xv

Part One: **The rural-urban fringe** — 1

1 Introduction — 3
- The edges of cities — 3
- Defining the rural-urban fringe — 5
- The physical fringe — 13
- Representing the fringe — 16
- A multifunctional fringe — 19
- The need for planning — 25
- Conclusions — 26

2 The making of the English rural-urban fringe — 27
- Introduction — 27
- Making the fringe — 28
- A unique landscape — 31
- Landscapes without community — 36
- Landscapes without politics — 38
- Transitional landscapes of production, redundancy and consumption — 40
- Conclusions — 44

Part Two: **Multiple fringes** — 47

3 A historic fringe — 49
- Introduction — 49
- Pre-modern archaeology — 52

	The industrial heritage and modern re-use	55
	The historic fringe and outdoor recreation	66
	Conclusions	69
4	**An aesthetic fringe**	**71**
	Introduction	71
	The physical landscape	72
	Representations of the fringe	76
	Changing landscapes	81
	Changing perceptions	85
	Conclusions	90
5	**An economic fringe**	**91**
	Introduction	91
	An economy in transition	92
	The new economy	94
	Economic survivors in the rural-urban fringe	105
	Conclusions	113
6	**A sociocultural fringe**	**115**
	Introduction	115
	Housing quality and residential experience at the edge	118
	Recreational experiences at the edge	125
	Accessing recreational experiences	131
	Conclusions	137
7	**An ecological fringe**	**138**
	Introduction	138
	Developmental versus ecological agendas	140
	Accommodating the ecological fringe	145
	Conclusions	156
Part Three: Planning on the edge		**159**
8	**Land-use planning and containment**	**161**
	Introduction	161
	Land-use planning and the fringe	162
	Containment and the fringe	164
	Beyond the green belt	167
	Alternative strategies: wedges and gaps	169
	Green belts: Is re-invention possible?	173
	Conclusions	179
9	**Planning reform and the spatial agenda**	**181**
	Introduction	181
	Spatial planning at the edge	182

Planning reform and the rural-urban fringe	187
Planning tools at the edge	196
Conclusions	200

10 Conclusions 201
Introduction 201
Delivering change – where necessary 203
Critical choices for the rural-urban fringe 211

References 214
Index 226

List of figures

1.1	Essential service functions	6
1.2	Warehousing, offices and allotments	7
1.3	Farmland: often scruffy and degraded	8
1.4	Warehousing at the urban edge	17
1.5	Discount fashion store	18
1.6	Colliers Moss Common	23
1.7	Wellesbourne and its airfield	24
2.1	Motorways truncating the landscape	30
2.2	Transition from 'rural' to more 'urban' land uses	32
2.3	A landscape fragmented by roads	37
2.4	Out-of-town shopping	41
2.5	Science park in its landscaped setting	44
3.1	The fringe is rich in industrial archaeology	51
3.2	Redundant industries	56
3.3	The industrial legacy	56
3.4	Derelict industries and formal leisure	58
3.5	Small general aviation airfield	61
3.6	Quarrying at the urban edge	65
4.1	An untidy landscape	76
4.2	Motorways and power stations	77
4.3	More out-of-town shopping	79
4.4	Power station – negative connotations?	87
4.5	Wind turbines – positive connotations?	87
4.6	Pylons	89
5.1	Major motorways and roads in the north west of England	95
5.2	Surface car parking associated with retail	97
5.3	Bristol Parkway	98
5.4	Yet more out-of-town shopping	101
5.5	Science park	104
5.6	Fly-tipping	106
5.7	Farming with airfield beyond	108

5.8	Mineral extraction	110
6.1	Golf course	117
6.2	Housing, allotments and canal	119
6.3	BedZED, Hackbridge, South London	122
6.4	Sketch showing Hockerton Energy Village in its 'ecological setting'	123
6.5	Recreation, canals and a potential 'greenway'	126
6.6	The fringe can have a lawless appeal	130
6.7	Cycling in the fringe	134
7.1	A landscape of 'unnatural' beauty	142
7.2	Wetlands	144
7.3	A landscape serving the city	150
9.1	Principles and outcomes – spatial planning at the rural-urban fringe	184

List of tables

1.1	Terminology, and the frequency of its use, in defining the fringe	9
1.2	Major land-use categories in the rural-urban fringe	13
6.1	Recreational land in the fringe	127
7.1	Landscape protection and 'environmental regeneration' tools	147
7.2	Corporate Social Responsibility – key themes	152
8.1	Green belt: an agenda for modernization	175
9.1	Principles for policy intervention	186
9.2	Past problems, planning reform and the rural-urban fringe	188

Glossary of acronyms

AA	Automobile Association
BAP	Biodiversity Action Plan
BEN	Black Environment Network
BSE	Bovine spongiform encephalopathy
CA	Countryside Agency
CAP	Common Agricultural Policy
CBI	Confederation of British Industry
CC	Countryside Commission
CF(P)	Community Forest (Programme)
CLA	Country Land and Business Association
CPO	Compulsory Purchase Order
CPRE	Campaign to Protect Rural England
CSR	Corporate Social Responsibility
CURE	Centre for Urban and Regional Ecology
DEFRA	Department of the Environment, Food and Rural Affairs
DETR	Department of the Environment, Transport and the Regions
DIY	Do it yourself
DoE	Department of the Environment
DTLR	Department for Transport, Local Government and the Regions
ERDP	England Rural Development Programme
FMV	Forest of Marston Vale
FoE	Friends of the Earth
GNP	Gross national product
HMRI	Housing Market Renewal Initiative
LDD	Local Development Document
LDF	Local Development Framework
LSP	Local Strategic Partnership
MOL	Metropolitan Open Land
MSW	Municipal Solid Waste
NIMBY	Not In My Back Yard
ODPM	Office of the Deputy Prime Minister

ONS	Office for National Statistics
PDG	Planning Delivery Grant
PDL	Previously Developed Land
POS	Public Open Space
PPG	Planning Policy Guidance (Note)
PPS	Planning Policy Statement
REACT	Regeneration through Environmental Action
RICS	Royal Institution of Chartered Surveyors
RSS	Regional Spatial Strategy
RTPI	Royal Town Planning Institute
SEERA	South East England Regional Assembly
SNCI	Site of Nature Conservation Interest
SSSI	Site of Special Scientific Interest
TCPA	Town and Country Planning Association
TIF	Total Income from Farming
UFAP	Urban Fringe Action Plan
UTF	Urban Task Force
WWF	World Wildlife Fund/World Wide Fund for Nature

Preface

Planning on the Edge – an examination of the context for planning and management at the edge of towns and cities – draws on research undertaken between 2003 and 2004, when a team from University College London were asked to look at the future of planning and land use management at the rural-urban fringe. This is often seen as a degraded area, fit only for locating sewage works, essential service functions and other, less than neighbourly, uses. However, the fringe is also a dynamic arena where a range of 'urban uses' – including out-of-town retail, decentralized office space, leisure facilities – collide with 'rural uses' including farming and forestry, outdoor recreation, and conservation. It has been called 'planning's last frontier' (Griffiths, 1994), an area that has slipped through the gap between town and country, and in urgent need of attention from policy makers. It has been estimated that more than a tenth of the land mass of the UK comprises rural-urban fringe: an area sometimes viewed as a potential resource for leisure, conservation, economic development, green energy production and so forth. This book is divided into three parts: what the rural-urban fringe is and what has underpinned its particular development; the context for planning and other forms of policy intervention at the fringe; and how we might, in the future, more effectively manage and plan the edges of towns and cities. The central aim of the book is to fill an important gap in the land use and spatial planning literature, examining in detail the context for planning in this 'no-man's land' or, as Augé (1995) might label it, this 'non-place' around towns and cities. We present problems and evaluate potential solutions, suggesting how planning might approach the 'problem' represented by the rural-urban fringe. Much of the analysis will focus on England, though a key objective is to draw out generic principles and ideas that are applicable to fringe areas around the globe.

The structure of Part Two of this book borrows from Brandt *et al.*'s (2000) application of the term 'multifunctionality' to landscape planning and management. Part Two is divided into five chapters that seek to understand, sequentially, the 'historic fringe' (Chapter 3); the 'aesthetic fringe' (4); the 'economic fringe' (5); the 'sociocultural fringe' (6); and the 'ecological fringe' (7).

A key line of argument in the book is that the rural-urban fringe is more than the sum of its parts: it is not merely an assemblage of particular land uses, or a transitional zone. Rather, it is a functioning landscape with unique attributes. The concept of 'multifunctionality' provides a means of organizing chapters (and reflecting a diversity of land uses and activities found in fringes), while also alluding to the complex dynamics of fringe areas. Arguably, past planning at the edges of towns and cities has been ignorant of this complexity, choosing simply to contain growth and turn a blind eye to fringe areas: planning at the fringe – in the form of green belt and broader containment strategies – has not been *for* the fringe. In the final part of the book, attention turns to past and future planning policy and practice, contrasting land-use planning and containment strategies with an emergent 'spatial' and more strategic planning model, which holds out the promise of more integrated and positive planning on the edge in the coming century. However, the book is concerned not only with the way in which the fringe is planned, but also with the degree of physical planning that is needed at the rural-urban fringe. Much of the 'potential' – aesthetic, ecological, social and economic – that we attribute to the fringe is a product of planners and policy makers ignoring these areas: this is an organic landscape that lacks the strict spatial organization of many urban areas. Hence, we ask whether a switch to more formalized planning at the fringe is the right solution for this landscape.

We are indebted to the Countryside Agency for supporting two projects looking at the rural-urban fringe during the past two years. We are particularly grateful to Christine Tudor and Andy Gale, for giving us free rein to interpret their requirements, and regularly step outside the confines of our original remit. Christine's enthusiasm for everything 'fringe' really provided the impetus to extend the original projects into this current volume. Thanks are also due to Marion Shoard and Richard Oades who worked on parts of the first project in 2003 and to a number of researchers who assisted with the policy reviews that formed the backbone of the original studies. In particular, Anna Richards, Jonathan Freeman, Emmanuel Mutale, Sibylla Wood and Joao Mourato all made important contributions. We are also grateful to Frazer Osment and LDA Design (Exeter) for their contribution to our work, which has informed some of our thinking on fringe landscapes, on permeability and on how access to the fringe might be widened. However, any faults or omissions that remain are entirely our responsibility. A final, but very special, thanks must go the office staff at the Bartlett School of Planning at UCL – to Lisa Fernand, Judith Hillmore and Laura Male – without whose brilliant administrative skills and immense patience, no-one in the School would be able to spend any time on research or dissemination whatsoever.

Part One
The rural-urban fringe

1

Introduction

The edges of cities

> Like the poor, the rural-urban fringe we have always had with us. Certainly, it has been with us since civilization first emerged over 6,000 years ago, and settlements gradually began to expand in function and size at the expense of the rural sector.
>
> (Thomas, 1990: 131)

There is nothing new in our concern for the rural-urban fringe: as Thomas points out, it has always been with us and is, quite obviously, an integral part of any urban system. Morphological studies of cities carried out in the 1920s reflected some early concern for the edge (Burgess, 1968 [1925]; Douglass, 1925), especially in the United States where increased car use and the development of freeways seemed set to obliterate any notion of a definable edge. In post-war Britain, a call for urban containment, delivered eventually in the form of statutory green belts, led to a paradox: an acknowledgement of the importance of fringes, but also a belief that these extraneous areas required relatively little positive planning. The nature of edges – in Britain at least – became assumed, but undefined. Green belt designation gave them a function, and seemed to put paid to any further debate over their future use or development.

But the urban edge has proven resistant to tight regulation: though many forms of economic development, and housing, have been prevented from sprawling beyond urban boundaries, the processes of change have not been halted. Within the urban shadow, farming has taken on special characteristics, becoming fragmented and struggling to tap into nearby urban markets; service functions (aspects of the 'urban dowry' – see Chapter 2) have been pushed to the fringe; illegal activities – including fly-tipping – have become common; and the fringe has become, for many, an expression of the tension between town and country. Indeed, Wibberley's (1959) study of agriculture and urban growth highlighted continued pressure on fringe areas, and Pahl's (1965) examination of Hertfordshire's metropolitan fringe revealed a complex rural-urban dynamic. Therefore, not content to entirely ignore the edge of cities, numerous writers have sought to understand the nature and role of rural-urban fringes, both as a component of the countryside, and as an integral part of towns (Pryor, 1968; Thomas, 1990; Whitehand, 1988). Elson's (1986) study of conflict mediation at the fringe – focusing on London's green belt – remains one of the most significant studies of planning on the edge. But despite this apparent concern for the rural-urban fringe – as a focus of analysis – Griffiths (1994: 14) has called the fringe 'planning's last frontier', arguing that areas abutting towns and cities have been largely neglected by land-use planning and by those agencies, public and private, with direct or indirect planning responsibilities. (This neglect, we argue in the next chapter, has been largely responsible for creating the contemporary fringe.) There are also persistent difficulties in defining exactly what we mean by the 'urban edge' or 'fringe'. Indeed, very recently a study from the University of Manchester (CURE, 2002: 18) has pointed out that if the fringe is planning's last frontier, it is a frontier without clearly defined borders. Common sense dictates that the fringe is something at the edge, bordering built-up areas on the one hand, and more open countryside on the other. But for some writers the fringe is merely the countryside 'around' towns and cities, not penetrating into the urban, but extraneous to it. For others, it is a zone of 'transition' that begins by being predominantly urban and ends up as mainly rural: that is, in terms of the mix of land uses, activities and the density of development.

But even in the absence of a clear definition of what the fringe actually is, there is still some agreement that fringe areas – in the UK and further afield – are poorly planned and managed. In some instances, they are merely ignored. In Britain, the former Department of Transport, Local Government and the Regions (DTLR) has pointed to the fact that land abutting many urban areas is frequently unkempt and a focus for unneighbourly functions, which are often pushed to the edge of towns, and away from people, in the hope of avoiding adverse public reaction, or to protect public health (DTLR, 2001a: Para. 3.24). Hence, waste disposal, car-breaking and sewage treatment facilities are regular features of the fringe. The fringe also faces considerable development pressure both from outward urban expansion and from a desire to direct particular types of development away from built-up areas to locations where land is cheaper and there is more room for future growth. Warehousing and distribution centres gravitate to arterial road junctions; retail and business parks exploit lower land prices and the abundance of open developable space; and a range of other land uses – commercial, leisure and so forth – are all attracted by these same attributes.

There are two major arguments that we will explore in this book: first, that there is a need for more effective spatial planning – including master planning and landscape management – around Britain's towns and cities. The second is that the fringe is not the product of conventional post-war planning, but has been created by more organic processes. The relative neglect of the fringe – compared to the attention lavished on other areas – has been a key driver 'making' this landscape (see Chapter 2). This means that the 'potential' that many agencies and observers now attribute to the fringe is unplanned: this begs the question as to whether the fringe really needs conventional planning, or whether such intervention would simply urbanize or 'countrify' the fringe, eliminating its uniqueness and nullifying its potential. Whichever path is deemed best, it is important to understand the special characteristics, needs and challenges of fringe areas. Moving to some conception of the rural-urban fringe would represent a key step in achieving such an understanding. From this platform, it might then be possible to develop appropriate management or planning strategies for this landscape. In the remainder of this opening chapter, we seek a preliminary understanding of the fringe and urban edge. We use 'fringe' to describe the landscape extending beyond the built-up area, and 'edge' as that zone close to what might be recognized as more conventionally urban. This discussion is built upon in Chapter 2 and also in the second part of the book.

Defining the rural-urban fringe

While we intend to explore the nature of the fringe throughout this book, and avoid simple definitions, it is perhaps useful at the outset to present a basic view of what the fringe is. In 2002, England's Countryside Agency defined the rural-urban fringe as 'that zone of transition which begins with the edge of the fully built up urban area and becomes progressively more rural whilst still remaining a clear mix of urban and rural land uses and influences before giving way to the wider countryside' (Countryside Agency, 2002b: no page number). Although useful by virtue of its simplicity – and this is precisely what is needed at this point – this definition is deficient in one particular regard: it fails to recognize the uniqueness of the fringe. We would suggest that the fringe is not merely a zone of transition stretching beyond the boundaries of the built-up area, with urban uses gradually giving way to rural uses and vice versa. This view of the fringe does have some validity: urban uses, including essential service functions, may in many instances be more concentrated close to the urban edge. But these are not standard urban land uses: rather they are peculiar to the fringe. They often serve the city – supplying energy or drinking water, or dealing with domestic sewage or commercial waste – but the desire to keep them separate from residential functions can mean that the fringe becomes the location of choice for such uses. It is also the case that uses requiring cheap space and access to arterial and orbital road networks – office or retail parks, light manufacturing, warehousing and distribution – may gravitate to the fringe (Figure 1.1). The roads themselves, along with railway marshalling yards and airports, also come to characterize this landscape. So on the one hand, the fringe does contain 'urban' uses that thin out away from the built-up area. Hence the transitional model has some currency. But on the other hand, these are not regular urban uses; rather, they are specific to the fringe. Instead of labelling them 'urban', it might be more useful and accurate to call them fringe land uses, hence moving to a functional definition of the fringe.

We enter into a much more detailed discussion of the nature of the rural-urban fringe in the next chapter and in the second part of this book, but it is

1.1 Essential service functions (Marion Shoard).

perhaps useful here to set out some of the land uses and activities that are common to different fringes. These will certainly include the service functions such as sewage works, rubbish tips, car-breaking yards, gas-holders, railway marshalling yards, motorway interchanges (see Figure 1.1). Commercial recreation facilities, such as go-kart or quad-bike tracks, golf courses, private fishing and water-sports lakes, nurseries, garden centres, football stadiums and golf courses are also regular features of the fringe. Such locations are also felt to be ideal for noisy or unsociable but non-recreational uses that depend on the presence of a large urban population, such as catteries and kennels. The same is true, at least in part, for travellers' encampments and caravan sites. In the future, more immigration and asylum reception centres may open up in the fringe, particularly in southeast England. In the past, mental hospitals – especially red-brick Victorian institutions – could be found on the edges of many towns and cities although a great many of these have closed in recent years as a result of the 'Care in the Community' programme, with the sites redeveloped mainly for residential or commercial use.

The proliferation of certain types of retail establishments – notably farm shops, nurseries, garden centres, superstores, retail complexes and shopping malls – is another theme picked up again in Chapter 2. The need for space – and low rents – is also the rationale for some factories, offices, business parks, warehousing (see Figure 1.2) to locate on the edge. Access to road infrastructure often means that these same uses are associated with or located on distribution parks. Cheap land can also provide the incentive for some formerly 'urban uses' to relocate to the fringe. For example, a growing number of educational institutions and district hospitals – which have sold off valuable city-centre sites – may move to the fringe to exploit the availability of cheaper land, affording them more room for expansion. We have already suggested that the fringe falls into the 'urban shadow', suggesting that it is peri-urban but strongly influenced by urban pressure and process. But there is an alternative view: that the fringe is peri-rural and within the rural rather than within the urban. Hence, urban encroachments transform areas of former countryside into fringe. Indeed, Audirac (1999) has distinguished between 'rural-urban' and 'urban-rural', suggesting that 'rural-urban' points to a dominance of 'rural change' in making the fringe, while the 'expansion of urban structure and function' are more important in the 'urban-rural fringe' (ibid.: 11). We use the term 'rural-urban fringe' in a neutral (alphabetical) sense, first because we believe that areas are subject to a complex array of processes, making it difficult to quantify the degree to which 'rural' factors have overridden 'urban' factors or vice versa, and secondly, because this is not a particularly useful debate. Areas that were previously open and perhaps agricultural have certainly been lost to urban uses and the fringe has been made by an array of processes, some of which might be described as 'urban'. But taking a view as to whether the fringe is within the urban or within the rural is perhaps regressive, if we wish to see the fringe as something unique and peculiar: the fringe is not merely corrupted countryside or low-density urban space.

That said, it is home to a range of additional uses that might be viewed as more rural than urban. Farmland on the edge of cities can often be scruffy or under-managed, for reasons that are explored in Chapter 2 and elsewhere in this book. Agricultural land may be intentionally degraded in the hope that planning authorities will wish to rid the area of an eyesore, expand settlement boundaries, and extend planning permission for new development in the next round of development planning. Alternatively, vandalism and illegal fly-tipping on what people consider to be 'waste land' may accentuate the scruffiness of the landscape. Yet another image of farmland at the fringe is one of well

1.2 Warehousing, offices and allotments (Marion Shoard).

managed horticulture, with farmers engaging in direct marketing of their produce – via farm shops – to the nearby urban populace. There is a view, explored later in this book, that farming cannot coexist with other more profitable land uses at the edges of towns and cities. Though this is sometimes the case, it is not universally so. Equestrian centres are another typical feature of the English rural-urban fringe, with portions of land given over to horse-grazing, training tracks and stables. So-called 'horsiculture' may become a key factor bringing middle and upper-middle classes into the fringe for recreation, for summer evenings or weekends of riding. Some country parklands survive at the fringe – Epping Forest to the east of London is one prominent example – which were once privately owned hunting estates but have today been opened up to public access. Country parklands sometimes highlight the potential of the fringe to play a broader recreational role; unlike other parts of the fringe, parklands are magnets for people. But elsewhere, difficulties with access, landscape degradation, noise and pollution, may limit recreational appeal. And finally, much of the remaining land area of the rural-urban fringe might fall under the label 'unkempt rough land', which may include land that cannot be developed because it is too steep or too contaminated; it may also cover land that has lost its developed function and has fallen into disuse or it may be awaiting development (Figure 1.3). Some ex-military sites might come into this category. But clearly, this is a grouping that includes land areas with very different characteristics and future potential: though contaminated land might be cleaned and reclaimed for economic uses, steep hillsides may provide a physical boundary to urban expansion, representing an urban edge that is not a fringe in the sense that we have employed thus far in this chapter. Indeed, fringes may be constrained by topography and by natural barriers. Hence, if a town is bounded by sea or mountains to the south, then fringe functions may

1.3 Farmland: often scruffy and degraded (Marion Shoard).

Defining the rural-urban fringe 9

be concentrated in the north or the east. If steep hills surround large parts of a settlement – as is the case in Bath – then 'ribbon' fringe may extend out next to a major road or along a valley that may afford the only means of access. Hence, fringes are rarely uniform girdles enveloping a settlement: segments of an individual fringe may well vary in thickness, density, content and form.

This is very much an Anglo-centric view of the fringe and its various ingredients, though worldwide there is some agreement that fringe areas are distinguished by their land use mix. Table 1.1 summarizes some additional features defining fringe areas, drawing on a number of different studies. There is agreement that fringe areas can be found on the edges of built-up areas (some of the studies examined did not make this point explicitly, but it was clearly implied). Secondly, patterns of land use were viewed as a defining feature in the majority of studies, with many authors referring to the types of uses noted above. Slightly fewer studies defined the fringe as an area of transition, and indeed, although the fringe may not be merely a transitional

Table 1.1 Terminology, and the frequency of its use, in defining the fringe

	Countryside Agency, UK (2002b)	CURE (2002)	Statistics Canada (2001)	Heimlich and Anderson USA (2001)	Hite USA (1998)	Bunker and Holloway Australia (2001)	ODPM UK (2001)	ODPM UK (2002a)	ReUrbA, UK/Netherlands (2001)	Broughton, UK (1996)	Shoard, UK (1999)	Urban Planning Act, Japan (1968)	Foot, Italy (2000)	Times cited
Edge of built-up area	X	X				X	X	X	X		X	X		8
Land use	X	X					X	X	X	X	X			7
Transition zone	X			X			X	X	X					5
Within built-up area		X					X	X				X		4
Metropolitan area			X	X								X	X	4
Within rural	X		X	X										3
Urban meets rural	X							X	X					3
Pressure zone		X						X			X			3
Population			X	X										2
Development										X			X	2
Policy	X												X	2
Statutory protection	X												X	2
Undeveloped land			X				X							2
Competition									X				X	2
Conflict										X			X	2
Urban 'shadow'			X											1
Spatial economy				X										1
Accessibility						X								1
Landscape											X			1
Green wedges	X													1
Way-of-life												X		1
Activities												X		1

landscape, it does display this characteristic in many instances. There is some disagreement as to whether fringes are integral to the urban, part of the rural, or something in between (a debate which we introduced above). Certainly they are areas of potential pressure, sometimes subject to particular policy restrictions (including green belt), have perceived development potential, and fall within an 'urban shadow': a status generating competition and conflict.

The consensus is that fringes are pressured landscapes, occupying a precarious position between town and country; they are characterized by particular patterns of land use. This can be illustrated using the case of England, but the English view of the fringe appears to have wider currency.

Of course, the suggestion that the fringe is a fringe by virtue of its proximity to a built-up area (and between town and country) is perhaps an obvious one. But again, the rural-urban fringe may not simply comprise a neat belt encircling a town or city. Sometimes, the belt is broken by physical barriers (as mentioned earlier); in other instances, fringes may penetrate into the urban core, perhaps following a natural feature such as a river or a man-made one such as a railway. Therefore, low-density light industry or warehousing on a river's floodplain may be labelled fringe development; likewise, old railway buildings on land adjacent to a mainline penetrating into a central urban core, today converted into business workshops, might also be considered fringe, particularly if some units remain disused and the land takes on the appearance of 'edgeland'. The characteristics of fringe areas will also vary according to distance from the urban edge.

The Centre for Urban and Regional Ecology (CURE, 2002: 18) has divided the 'countryside around towns' into 'urban edge', 'inner fringe' and 'outer fringe', each with distinct land use characteristics. In Australia, Bunker and Holloway (2001: 2) have suggested that all fringes possess an 'inner edge' and an outer 'periphery'. This subdivision of the fringe into a fixed typology is not a line of analysis that we wish to pursue in this book, partly because it leads back to the 'transitional' model examined briefly earlier. The inner edge may have a certain collection and density of urban uses; the periphery may have lower density horticultural uses. In the same way that the countryside or towns may have their internal configuration categorized and labelled, so too may the fringe. Remoter countryside will be less accessible to major urban centres; it may have a more fragile, seasonal economy; its housing market will be markedly different from that of the accessible countryside, to which retiring households and commuters flock. Likewise in cities, central business districts will have less residential space than suburbs; and suburbs will have less fashion retail and cultural land uses than the inner core. But all types of rural and all forms of urban have characteristic land uses and activities: it is taken for granted that form and content will differ, and yet the general labels retain currency.

'Fringe' has always been a more fragile label: areas of fringe cannot be delineated merely by economic activity and land use, or by density of physical development. Fringe is a differentiated form of landscape, though characterized by the assemblage of land uses that we have set out on the previous pages. Land use mix has been held up as the defining characteristic of the fringe; however, it is extremely difficult to accurately quantify patterns of land use in Britain's rural-urban fringe. A paucity of reliable data is largely the result of the lack of current agreement over what actually constitutes fringe land (but, then, without the data in the first place, it may prove impossible to realize a workable definition). And so, despite all the discussion over what the fringe is, we are left with no clear view. This has resulted in the proliferation of a wide range of assumptions: that the fringe is a fixed halo extending out from urban boundaries, delaying the onset of real countryside; the fringe is a no-man's land weaving in and out of the city, but roughly encircling it; the fringe can be measured and zoned; the fringe is intangible and defies strict definition. The lack of data has also resulted in a tendency to 'characterize' the fringe in both policy and media representation. Through characterization, the fringe becomes a 'landscape' presenting a range of challenges to policy makers, though the nature of these challenges and how they should be addressed are often left open to interpretation. For instance, though the recent 'England Rural Development Programme (ERDP) Annual Report to the European Commission' (2003) says little about the fringe from a national perspective, some of the regional chapters focus in detail on the fringe and its characteristics. The report notes that England's north west Region

> can be described as having three distinct rural areas, each with their own rural development issues. The areas are: the uplands of Cumbria and the Pennines, the lowland plains of West Lancashire and Cheshire and the urban fringes of Greater Manchester, Merseyside and other industrial towns.

On the same page, it is noted that the fringe has competing demands for land and faces 'many threats and opportunities', stemming from its proximity to large urban populations. Apparently, the 'challenge' in the fringe is to better integrate urban and rural communities while recognizing the importance of fringe 'wetlands, lowland raised bogs and woodlands'. The fringe also has large areas of derelict and degraded land as well as community forests that provide a focus for employment and recreation. Horticulture is a particular feature of fringe landscapes.

Similar messages emerge from the other regional reports contained within the same document, which all point to: a degree of dereliction and vacancy, of both open land and buildings; important habitats that are often threatened by urban encroachment; woodlands and other types of habitat that offer a potential focus for tourism; specific forms of agriculture (e.g. horticulture); and a coming together of urban and rural communities. This type of characterization is common in policy literature: it offers some clues as to what the fringe is, but makes no explicit assertions. Much is left open. Therefore a picture of the fringe needs to be built up, beginning with a view of the underlying forces of landscape change that have made the fringe. This is a process that we begin in the next chapter and continue throughout the remainder of this book.

We are not alone in trying to understand and articulate the nature of fringe landscapes. Much has been written on the fringe, in both the developed and the developing world. In the US, there has been heightened interest in recent years in urban encroachment onto agricultural land and the processes creating fringe landscapes that extend beyond but are demonstrably different from suburban landscapes (see Heimlich and Anderson, 2001; Carrión Flores and Irwin, 2004). In the developing world, there is increasing concern for the expansion of urban populations onto the fringe and the implications for agriculture versus the problems of supplying basic services (see Nkambwe, 2003). Planning and the processes creating uniqueness at the fringe are a recurring theme: Carruthers (2003), for example, has focused attention on political fragmentation at the urban edge and the impact that this may have on creating the peculiarities of fringe landscapes. This is an issue that we return to in Chapter 2. Bunker and Houston (2003) have examined similar issues at the fringes of Sydney and Adelaide, raising questions over the future of governance at the rural-urban fringe and arguing that land-use planning might give way to softer management approaches (ibid.: 303). This is not dissimilar from the line of argument that we follow towards the end of this book.

There has also been a concern for the historical development of fringes (see Thomas, 1990), with Whitehand and Morton (2003) looking specifically at Edwardian fringe belts in England associated with 'the hiatus in house building before, during and following the First World War' (ibid.: 823). In these belts *within* the city, Whitehand and Morton note an historical assemblage of informal open space, allotments, playing fields, golf courses, parkland, sports stadiums, institutional land uses, utilities, commercial development and water (e.g. reservoirs). Fringes have developed, and some have been consumed by subsequent urban growth. Returning to the quotation that we began this introduction with, Thomas (1990) has noted that fringes have always been a feature of human settlement; as settlements have grown, fringes have been lost, but ultimately remade. Fringes are dynamic landscapes in the sense that they face continual pressure and survive as a melting pot between the urban and the rural, between the pressure to grow and the desire to protect. In contrast, the pace of physical change in the wider countryside may be far slower, though social and economic shifts will of course remake the landscape over time. In cities, there is a different kind of social and economic dynamism, with capital constantly relocating, causing a hollowing out of urban cores, a pressure for public sector intervention, and rapid social reconfigurations. As we suggest in the next chapter, the fringe is subject to the broader national and global trends affecting all places; but these trends often work the fringe in unique ways, producing particular physical outcomes.

We are following a well trodden path, but aim to examine the fringe in a comprehensive way, arguing that this is a unique and a special landscape. There are two more themes that recur throughout this book. The first is that the fringe – like other landscapes – is constructed through representation as well as physicality. This is an issue that we take up at length in Chapter 4. The fringe is imagined as much as it is physically understood, and therefore a major challenge is to understand the fringe and its representation in the modern psyche. The second is that the fringe is an unplanned landscape (in many respects), with capital 'making' the fringe by pursuing opportunity in a largely unfettered manner. This has created a diverse landscape, rich in historical association, in ecological diversity, in recreational opportunity, in social capital and in aesthetic variety. This leads to the question of how much physical planning is needed at the fringe; will too much merely act to urbanize the fringe, extending cities and obliterating the potential that many observers and agencies are now attributing to this landscape? These are themes and questions that we pursue at length throughout this text, and which are now introduced in more detail.

The physical fringe

We have already sketched out the uses that locate in the fringe as a result of abundant space, cheap land and good access; and the uses that are dispatched to the fringe to avoid conflicts with people, and with less noisy or less dirty uses. A range of processes – historical, contemporary, economic, social and political – have made the fringe in a sense that we explain more fully in the next chapter. For the moment, it is perhaps worth adding some detail to our sketch of the fringe, and focusing on the mix of contemporary land uses. In a study for the Countryside Agency (Gallent *et al.*, 2003a), we were asked to examine the characteristics of nine major land-use classes in the fringe. The objective of this work was to show how different land uses might coexist, and might be managed using softer management approaches (than simply land-use planning) of the type noted by Bunker and Houston (2003) in an Australian context. The issue of bringing uses together and combating the 'fragmentation' of land management at the fringe (Carruthers, 2003) through a multifunctional framework based on some of the concepts is introduced in Chapter 3. Drawing on the observations from our earlier work, Table 1.2

Table 1.2 Major land-use categories in the rural-urban fringe

Waste management facilities	*Mineral extraction*
New landfill and ex-minerals sites (i.e. former quarries or surface mining facilities) are often found in the rural-urban fringe. New processing facilities may be associated with the clean-up of heavy industry on the edges of towns and cities, or may utilize redundant sites or buildings that are to be found in the fringe. Similarly, modern facilities for dealing with the recycling of waste may replace older incinerators (and so forth) that have traditionally been viewed as un-neighbourly and therefore pushed into the fringe. There is an ongoing modernization of waste facilities within the UK's fringe, and even a resurgence of incineration associated with energy production.	Some aggregate and sand extraction may occur in the rural-urban fringe, especially where soft materials are extracted from nearby floodplains: an often cited example is that of the Jubilee River near Maidenhead, which is an artificial channel providing flood relief for surrounding towns. However, a key feature of many fringes is the legacy of abandoned workings associated with a heavy industrial past. These old workings sometimes provide recreational or waste management opportunities (see left).

Table 1.2 continued

Energy production and distribution

Major energy production facilities may locate in the fringe away from, but also in close proximity to urban populations. Transformer substations are also a feature of many fringes and cluster with other utility land uses, often in the inner fringe zone. However, it is also the case that newer forms of energy production are found in the fringe. It was noted above that modern incineration plants may be associated with energy production. Proximity to cities makes these locations ideal for the composting of urban wastes, the harvesting of wind energies (where feasible) and the production of bio-fuels associated with urban fringe agriculture. Lower-density housing development in the fringe and in green belts introduces opportunities for solar energy production and for grey water recycling. Similarly, warehousing can also be associated with solar energy, generated from flat panels fixed to warehouse roofs.

Transport infrastructure

Orbital and arterial roads are an obvious feature of the fringe, and the role they have played in 'making' the fringe is discussed in Chapter 2. These are associated with junctions, roundabouts, service stations, park and ride schemes, retail and distribution parks and with a general decentralization of some office and retail space. Indeed, the density of road networks in the fringe may provide an impetus for the urbanization of the countryside around towns. Railways also transect the rural-urban fringe and many fringes now play host to 'parkway' stations located where roads and rail networks cross. Parkways are new stations that relieve pressure on urban terminals and provide opportunities for commuters to drive to a station, leave their car in sizeable car parks, and

Recreational land uses

A broad spectrum of recreational land uses locate in the fringe. These include parklands (discussed in the main text) associated with historic buildings and landed estates. Some allow free access while others remain private. Other forms of land use with recreational potential include woodlands, allotments, common land, golf courses, rivers and canals, and lakes and reservoirs. The potential of these land uses to provide realizable recreational opportunities will depend on existing rights of way, the negotiation of further access agreements, the establishment of community forests, the implementation of rights of way improvement plans, or the setting up of stewardship schemes. Many country parks, regional parks and greenways can also be found in the urban fringe. These may offer significant opportunities to open up the fringe to less formal recreation. It should also be noted that formal recreation – sports fields, running tracks and so forth – is also a feature of the fringe.

Commercial development

A wide range of commercial activities can be found in the fringe, especially retail parks, science parks, warehousing and light industrial land uses. Together with essential service functions, this range of very particular commercial development forms the full spectrum of major employment uses in the countryside around towns. Retail parks and major out-of-town shopping centres became a key focus of planning policy in the 1970s and 1980s, only to be rejected in the following decade as their full effects on town centres was realized (see Chapter 2). But today, despite PPS6's broad rejection of out-of-town retailing, it remains clear that certain forms of bulk retailing (hardware superstores, garden centres, specialist wholesalers etc.) are often more suited make the rest of their journey to work by train. As well as parkways, fringes may also be the location of freight and maintenance yards. Waterways (canals and navigable rivers) may also be found in the fringe though their potential lies more in recreation than in passenger or freight transport. Waterways may provide a focus for greenways – corridors that provide opportunities for walking and cycling. Returning to hard infrastructure, major airports and smaller aerodromes also locate on the edge of built-up areas and add to this unique balance of land uses found in the fringe.

Conservation sites and historical/ archaeological sites

Key amongst conservation sites in the rural-urban fringe are former industrial buildings that may help preserve a sense of historic association/continuity. Former mills, old airfield buildings, agricultural buildings of historic interest/ importance can all be found in the fringe. Indeed military establishments, abandoned after the ending of the Cold War may also be found in some fringes. Added to these are the historic parklands and houses that once stood in open countryside but now find themselves bordering a growing town or city. Many rich habitats are also found in the fringe – these may seem tacky or shabby at first glance, but are rich in biodiversity.

to fringe locations, largely because such operations require considerable expanses of floor space which would be prohibitively expensive in many centres. More specialist retailers (plumbers' / builders' merchants, large car showrooms etc.) often favour fringe locations for the same reasons. Light industry shares this need for space and the growth potential afforded by a non-centre location. Indeed, many light industrial uses once occupied prime commercial space in many town and city centres but eventually left because of spiralling costs and a lack of space. Loading and distribution is also more difficult in central locations, while the best fringe sites frequently have good access to radial and orbital roads, and therefore good access to customers. Warehousing and distribution favour out-of-town locations for much the same reasons: cost, space and access. And the more recent development of business and science parks share similar concerns and priorities.

Housing

Residential development may be an important element of fringe land use in some instances, though often the fringe is a landscape 'without community'. Between 1998 and 2001, 4 per cent of housing completions in England were in the green belt. Across all fringe areas, housing development tends to be lower density than elsewhere. In the past, it has straddled arterial roads to form ribbon development; today, it often concentrates around road interchanges and may be associated with office or retail development (forming mini 'edge cities'). There are concerns that residential development in the fringe fails to achieve quality in terms of either design or sustainability. This is the realm of the volume builder and of low density redbrick development occupied largely by car-dependent households. However, opportunities might be created to promote green housing in the green belt –

16 *Introduction*

Table 1.2 continued

and in other fringe locations – demonstrating that soulless boxes need not be the norm.

Farming and forestry
Agriculture and forestry are integral parts of the landscape at the fringe. Forestry may take the form of commercial woodland. There are also twelve 'Community Forests' on the edges of built-up areas in England. Farming at the fringe faces a range of challenges, not least from direct urban pressures such as housing and infrastructure development, and from indirect pressures including illegal fly-tipping. Some agricultural land at the fringe may be degraded because of these pressures. However, a location in the urban fringe may give some farms competitive advantage, especially when they exploit their proximity to large urban markets. Organic farms, in particular, may engage in short-chain product marketing or in direct marketing to consumers living in built-up areas. Other farms may diversify into activities that capture an urban market: they may, for instance, stock rare breeds and promote themselves as educational centres; others may sell 'activity weekends'. The market potential for such diversification is likely to be greater in the rural-urban fringe than in the wider countryside. Indeed, farming faces threats, but also enjoys particular opportunities by virtue of its location in the urban fringe.

---

summarizes the characteristics of key land-use categories found in England's fringes.

More is said on the nature of these land uses throughout this book, and especially in Part Two. Chapters on the historic fringe, the aesthetic fringe, the economic fringe, the social fringe and the ecological fringe reflect on patterns of land use and land use change, examining commercial activities, transport, agriculture, housing, heritage, recreation, and energy, waste and mineral extraction. All are given at least some attention in this book.

But the key purpose is not merely to describe – and therefore arrive at a descriptive characterization of – the fringe, but address major questions of action: to detail the required nature of planning at the edge. To achieve this, it is necessary to understand not only the physical fringe, but representations of the fringe (and how these might need to be challenged), together with the desired principles for future action (given the nature of the fringe as a dynamic, pressured and highly diverse but fragmented landscape) and the case for more conventional planning at the edges of towns and cities versus a softer approach, offering greater flexibility, which might challenge convention. We now look at each of these issues in turn: representations of the fringe; the principles for future action; and the case for alternative approaches.

Representing the fringe

The land uses commonly found at the rural-urban fringe combine to form a particular physicality: these are its constituent parts. But the landscape of the fringe is considerably greater than the sum of these parts: the edge of the city is commonly 'defined' through character and perception, which frequently results in negative representation. The fringe as necessarily functional, as chaotic, and as essential but ugly is often contrasted with the intrinsic charm of the countryside beyond. In Chapter 4 we examine the aesthetics of the fringe, explaining why the fringe is often held in low regard, why it fails to please conventional landscape tastes, and why it is often seen as the antithesis of a 'quality landscape' (Shoard, 2002: 121). Essentially, the functionality of the fringe, the pace of changes that can occur, and the negative associations of many of its land uses can add up to negative perception, which is subsequently amplified through negative representation. Many of the regular landscape features of the fringe – including power stations – might be viewed as icons of a vulnerable modern world, and of a functionalism that seems to challenge nature, polluting the landscape (Figure 1.4). Society has developed a technophobic view of the fringe, punctuated by essential but unattractive features. Thayer (2002) has illustrated that functionality has associated symbolism: green energy production is more likely to be viewed as a positive landscape attribute than facilities burning fossil fuel. Landscapes are the product not only of their 'physicality' – the roads, power-plants, warehouses, waste facilities, houses, etc. that create a particular vista – but of the meanings attached to these items. This is an important message for planning at the fringe, and one that we return to in Chapter 2.

We begin then with a negative perception of the fringe, as a landscape tolerated out of necessity, but certainly not celebrated. We demonstrate in the next chapter that this perception has been amplified – time and again –

1.4 *Warehousing at the urban edge, Shifnal, Shropshire (Countryside Agency, Rob Fraser).*

18 *Introduction*

through negative representation. Hollywood movies often attribute 'apparent lawlessness' (Shoard, 2002: 130) to the fringe. This is a landscape of abandoned factories, derelict buildings: in the film *Memento* (2000), it was seen as the ideal location for exchanging large sums of money for drugs, for committing murder, and ultimately for concealing bodies. A similar image emerges in *Fargo* (1995) when the fringe – ignored, anonymous and seldom visited – is the location of choice to hold a kidnap victim. The physicality of the fringe is a catalyst for representations that enforce prejudice: this is an additional problem for planners and policy makers who deem it necessary to re-plan and re-make the fringe. Returning more to the physicality, dense transport and energy infrastructure together with large-scale commercial developments such as out-of-town shopping centres and business parks have had a great impact on the contemporary urban-fringe landscape (Figure 1.5). They contribute to an anonymous feel, which approximates with Marc Augé's notion of 'non-place'. In 'Non-Places', Augé argues that 'super-modernity' is characterized by a type of place 'which cannot be defined as relational, or historical, or concerned with identity' (Augé, 1995: 77). The fringe may take on the appearance of a landscape littered with an assortment of disconnected uses, thrown onto the landscape with little forethought.

These are all issues that we return to periodically throughout this book. The fringe cannot be characterized by physical items alone; it has a place in the collective consciousness shared even by those who have spent little time exploring or studying its intricacies. It is also a landscape of representation. People may not pay regular visits to the edges of towns or cities, but they have

1.5 *Discount fashion store, Ross on Wye, Herefordshire (Countryside Agency, Rob Fraser).*

an instinctive perception of what might be found there: either an assemblage of urban activities pushed beyond the edge or a corruption of more traditional rural landscapes. The task for planning – if such a task exists – is perhaps two-fold: to deal with the physical challenge created by the interfacing of urban and rural uses; and to address concerns over image and place-making at the edge. These are issues returned to in Part Three.

A multifunctional fringe

Another key anchor for this book is the idea of multifunctionality. If we accept the need for some form of policy intervention at the fringe – either hard planning or softer management – then such intervention needs an underlying rationale and purpose. Perhaps this purpose should be to manage the fringe's inherent diversity and dynamism, and to realize the potential – for recreation, for economic development, for sustainable waste management, and for green energy production – that many lobbying groups claim are achievable. We have noted that the fringe is a 'functional' landscape (and derided by those who associate function with low landscape value); it also has multiple functions – essential services, an assortment of economic uses, recreational potential – that have been 'thrown onto the landscape' and remain disconnected, in part because of the political fragmentation of the fringe (responsibilities are split, as we will see in the next chapter). Given this context, the idea of 'multi-functionality' may have significant value in framing future interventions. For urban areas, 'mixed use' has become a goal of much contemporary spatial planning (i.e. planning with a broader set of socio-spatial objectives); but multifunctionality is a concept borrowed from agricultural economics and landscape science. It is not merely a goal or a labelled reality, but a guide for understanding how landscapes function (or how different functions may share space), why they may function badly, and how they might function better: a goal and a guide for action.

In recent years, the term 'multifunctionality' has become fixed in the nomenclature of planning. It has become synonymous with practices that avoid the compartmentalization of land use or simple land use planning solutions. It is part of the spatial planning agenda that is explored in greater depth towards the end of this book. We use the term here as a means of understanding planning's possible role in the rural-urban fringe: multifunctionality suggests how different activities might be beneficially combined; it shows how broader objectives can be achieved that need not be led solely by the public or private sector, or need not focus on single issues and problems. Again, the term is rooted in agricultural economics and refers to the way in which individual farm units may expand into new productive areas (e.g. from mono-functional cereal production into vegetables, livestock, fruit and so forth) or into pluriactivity or farm diversification strategies. More recently, it has been used to denote a new phase in farming production, stretching from 'productivism' through 'post-productivism' and now to 'multifunctionality', though the linearity of this progression has been challenged. Indeed, it is perhaps difficult to claim that farming has ever been 'post-productionist', and therefore multifunctionality becomes a means of describing the reality of some farming strategies, where landowners not only engage in primary production, but may allow recreation uses (for instance) on their land.

On the one hand, multifunctionality is concerned with modes (and varieties) of production, but on the other, it is about how land is used in a wider sense. The intensification of agricultural activity in the post-war period led to a reduction in farming practices that might be labelled multifunctional (the emphasis switched to intensive production). But it could be argued that the development of agri-environmental initiatives during the 1990s changed attitudes towards the productive potential of farm land, and showed how 'production' can give way to a greater range of functions, some of which continue to deliver a profit for farmers. However, this focus on agriculture and on a 'productionist' view of multifunctionality might be seen as rather narrow. If the functioning of farming landscapes can be conceptualized through multifunctionality, why can this same concept not be applied to non-farming landscapes?

Indeed, the desire to plan and manage landscapes more 'sustainably' has resulted in the adaptation of the concept, first as a means of understanding land use and landscape change, and secondly, as a conscious guide for planning and management. According to Brandt and Vejre (2003 [2000]: 2) 'the concept of multi-functionality is getting increasing attention not only in the landscape sciences but in society in general, since it seems to be an important aspect of . . . sustainable development'. In urban debates, particularly in relation to current initiatives to encourage an urban renaissance and compact city structures, the term has come to complement 'high-density mixed-use development'. In regions with a high population density such as the Netherlands, multifunctionality has been 'used as a planning concept, which addresses the planning challenge to concentrate and combine several socio-economic functions in the same area, so as to save scarce space and to exploit economies of synergy' (Rodenburg and Nijkamp, 2004: 274). The separation of urban functions can be seen partly as the legacy of industrialization, which separated live and work spaces (Priemus et al., 2004: 269), but more recently modernism has created a mono-functional planning paradigm in which zoning became the norm (Rodenburg and Nijkamp, 2004: 274). However, contemporary debates on how the built environment can be sustainable tend to stress that the way forward is to lessen the impact of the existing built form by enabling it to perform several simultaneous functions. Hence, multifunctionality can provide a guiding light – and a fairly simple principle – for planning and land-use management in all contexts. But at the fringe, it may prove particularly useful given the range of land uses and activities competing for space. The sharing of that space (with local partnerships orchestrating land-use efficiencies) seems like a sensible goal in this particular context. Wood and Ravetz (2000) have shown that the rural-urban fringe is a landscape with great potential for enhanced multifunctional land use. An integrated fringe strategy, they argue, could embrace:

> agriculture combined with woodlands, wildlife and smallholding;
> woodlands combined with leisure, education, wildlife, smallholding

and low-impact housing; leisure combined with education, smallholding, woodlands and wildlife. The often stark division between town and country needs to be broken down in creative fashion, one in which the former goal of land productivity is steered towards organic cultivation, woodland culture, horticulture and permaculture, all now seen as viable creators of employment and maintainers of diverse and abundant landscapes.

(Wood and Ravetz, 2000: 15–16)

Put very simply, multifunctionality appears to provide a set of rules guiding more sustainable planning and management of fringe areas. Brandt *et al.* (2000) have sought to distil the literature on multifunctionality (particularly that literature applied to landscape science) into a simple message: that spaces perform basic functions – economic, ecological, communitarian and aesthetic; and that they have historical associations. From this perspective, they suggest that planning can either reject or accept the 'spontaneous' interplay between functions. Rigid land-use planning has tended to edge towards rejection (promoting the separation of land uses); softer management approaches, however, might try to work with rather than against the grain of 'natural' process. They summarize and describe the basic functions in the following way (ibid.: 26):

1. 'Ecological functionality', meaning that landscape is 'an area for living' for both human and non-human life;
2. 'Economic functionality', with landscapes providing 'an area for production';
3. 'Sociocultural functionality', or 'an area for recreation and identification' with sociocultural attributes;
4. 'Historical functionality', or 'an area for settlement and identity' that offers a sense of sociocultural continuity;
5. 'Aesthetic functionality', with landscapes providing 'an area for experiences'.

According to the same authors, 'the future management of landscape must include some kind of multifunctionality in its approach . . . it is a task of spatial planning to assign function and future forms of function and use to land' (ibid.: 26). Landscape management should certainly avoid some of the practices identified by Ling *et al.* (2000) that have underpinned the regeneration of derelict industrial land. They observe that derelict land regeneration has usually resulted in the redevelopment of discrete sites for industrial, commercial or housing end uses rather than the holistic regeneration of a whole landscape involving the exploitation of its potential for the range of functions listed above. This spatial segregation, they suggest, has neglected the potential of such places as a habitat for wildlife in particular. Consideration of the land as separate parcels has been reinforced by institutional factors with various organizations involved in regeneration approaching their task from a single-use perspective. This is certainly true at the fringe, where Carruthers (2003: 475) points to political fragmentation as fundamental to the way policy interventions may create a splintered, disjointed landscape of disparate and disconnected uses.

But are there any examples of multifunctional planning in action that might guide our thinking on this particular issue? There are schemes and places that illustrate some of the features described above, but few that consciously adhere to these prescribed principles. In 1990, Groundwork St Helens, Knowlsey and Sefton were invited by British Coal to draw up a reclamation strategy for Colliers Moss Common (130 hectares of post-industrial land on the site of the former Bold Power Station on the edge of St Helens in Lancashire) (Figure 1.6). The site formerly comprised a colliery with associated spoil tips and the coal-fired power station, all to the south of a railway line. The power station closed in the 1980s, rendering the colliery redundant. The northern section of the site was previously known as Bold Moss. Because the site had previously been an important peat bog wetland (filled with colliery spoil and waste since the 1960s; Morgan, 2002: 4), it was felt that this rich ecology should be reclaimed. The aim was to establish a nature-rich urban common for local people. A Bold Moss Forum was established to provide a link with local users: a key aim was to integrate natural regeneration (a renewed and widened ecological functionality) with community involvement in the first instance, and longer-term sociocultural functionality. For example, 'natural variations in soil conditions produced a small-scale vegetation mosaic which requires gentle management suited to volunteer work' (see EcoRegen Web Site). Fifteen years after the project began, the site is widely used as an informal recreation resource by local residents. Arguably, by accident rather than by design, multifunctionality was achieved in a number of ways:

1. Some of the former colliery buildings were renovated for community businesses, maintaining the site's function as an area for production;
2. Spoil areas were developed into a community park emphasizing biodiversity. Figure 1.6 shows the edge of this new park, on the edge of existing housing, which previously looked onto spoil heaps that were out of bounds to local residents;
3. A motorcycle track and fisheries have been created. There are also new landscape features (including a Millennium Bridge joining the northern and southern sections of the site), and sculptures commissioned from local artists. The site thus provides for a wide range of recreational and cultural experiences;
4. Historical association: the site's industrial past has not been discarded, but rather it is celebrated and used to create a sense of place and identity. Some former colliery buildings have been retained and machinery positioned around the site to form points of interest in the landscape, mixing the wild with the humanized. In particular, old pit wheels have been re-used as footbridges.

The implicit multifunctional and multi-agency strategy for Colliers Moss has been viewed as highly innovative. Indeed, experience gained at this site provided a framework for the 'Changing Places' programme, 'an ecologically

A multifunctional fringe 23

1.6 Colliers Moss Common, St Helens, Merseyside (Groundwork Trust).

informed and community led programme of land regeneration' (Barton, 2000: 5) promoted by the Groundwork Trust together with a number of national and local partners (The Millennium Commission, English Partnerships, Welsh Development Agency, the European Union, the Countryside Agency and the Office of the Deputy Prime Minister). By 2001, a total of 21 projects had been undertaken employing the approach first used at Colliers Moss. Colliers Moss can be held up as an example of what Brandt and Vejre (2003 [2000]: 21) term 'real multifunctionality' with the integration of different uses within a single space.

The reality of economic restructuring and global trade have rendered many former industrial sites obsolete. Wholesale industrial redevelopment is rarely a viable option: multifunctional re-use, however, offers a combination of landscape improvement, sociocultural opportunity and new economic openings. None of these individual gains could have been achieved in isolation; all are closely linked. In this instance, an integrated strategy, aiming to achieve spatial integration and added value (across aesthetic, sociocultural, economic, ecological and historic dimensions) through a multi-agency and participatory approach was seen to offer clear rewards. In the penultimate chapter (9) of this book, we return to look at how such apparent successes might be replicated and built on the same implicit principles.

Colliers Moss is a single illustration, but it does highlight attempts to think in a more holistic way about the reclamation and management of rural-urban fringe sites. There are other examples of typical edge sites that seem to fit

24 *Introduction*

the model. For example, old military airfields – which we examine in more detail in Chapter 3 – have often been re-used in a variety of ways with different activities seemingly sharing space in efficient ways. Bell *et al.* (2001) have examined the legacy of old airfields in Suffolk, Warwickshire and Yorkshire. In Warwickshire, the field at Wellesbourne – which ceased its military function in 1964 – is located on the edge of the village of Wellesbourne (a settlement of close to 10,000 residents) and has gradually developed an important relationship with the local community: not only through historic association, but also through job creation, providing space for new housing, and creating opportunities for other economic and sociocultural activities. Figure 1.7 shows the location of the airfield adjacent to the village and also pinpoints some of the different uses. The airfield is located to the west of the village: it is home to a weekend market, motor sports, agriculture, motor vehicle storage, industrial storage, distribution and light manufacturing, an airfield museum and housing development (both new at the north east corner, and more established at the Dovehouse Estate). Many airfields have become multifunctional, usually for one of two reasons. First, old hangars provide ideal buildings for new industries, and closed runways provide great venues for motor sports, markets and car storage. And secondly, a desire to retain flying operations sometimes means that owners will seek to cross-subsidize flying by leasing space for other activities. Airfields illustrate the kinds of relationships that frequently develop between the legacy of the past and the needs of the present. Similar patterns of re-use could be observed at Colliers Moss, with the old colliery taking on a wider socio-economic function. An almost identical transformation has occurred at Wellesbourne, though this has been more

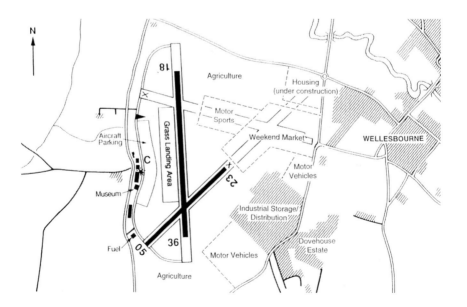

1.7 Wellesbourne and its airfield.

gradual and is the product of individuals grasping particular opportunities rather than any holistic vision. Nevertheless, both examples serve to demonstrate the nature of the historic fringe and how this may – either by fortune or by design – be linked to other fringe functions.

It is possible to make rather grand claims for 'multifunctionality': that it is the primary objective for all planning and land management; that it enables a clear conceptualization of the functioning of space; that it is the legitimate measure of success in all future policy intervention; that it is a paradigm for socio-spatial action and the reality of the world around us. But concepts are frequently rendered meaningless either by overuse or by misuse. 'Sustainability', for example, is one of those unfortunate concepts that suffer both abuses. Everything today that policy claims to aspire to, or that government hopes to achieve, must carry the label of sustainability: hence people will live in sustainable communities, use sustainable modes of transport, eat sustainably sourced food, sustainably dispose of their household waste, and generally aspire to be more sustainable. At the same time, we are told that 'balanced' communities are sustainable communities, that 'quality' public services equate with sustainable services, and that acting locally but thinking globally will mean that we have bought into this perfect paradigm.

Therefore, we will use multifunctionality in two rather modest ways in this book. First, the concept is used to frame discussions of planning strategies at the rural-urban fringe, mainly in later chapters, but also at some points in Part Two. Secondly, we use the term as a guide for pulling apart and reassembling land uses into a structure for this book which reflects the broad areas of concern. This second point is perhaps the most important. The structure of the second part of the book attempts to reflect what the fringe is and how it functions (therefore it shares multifunctionality's broadest aims). Chapters are presented which look at the historic fringe (3), the aesthetic fringe (4), the economic fringe (5), the sociocultural fringe (6) and the ecological fringe (7). In each chapter, an attempt is made – where relevant – to reveal how the landscape of the fringe functions in terms of how different land uses come together either in a state of conflict, or beneficially.

The need for planning

Issues of representation and multifunctionality provide two key anchors for this book; special attention is paid to issues of representation in Chapter 4 (looking at the 'aesthetic fringe') while multifunctionality provides a perspective on planning in the final chapters. But a critical point here is that the need for planning at the edge (or more planning, or more conventional planning) is not a given. We have already noted – and will build on this idea in later chapters – that the fringe might be described as an organic landscape, which has not been 'reined in' by planning to the same extent as cities (subject to various forms of intervention) or the wider countryside (subject to rigid land-use controls since 1947). There is a certain lawlessness associated with the fringe, extending to both how the fringe is used and how it has been managed. It has been ignored by policy makers, and has found few champions to question its development over the last half century. But that has not prevented it from developing a potential that is now widely recognized. However, potential can be defined in a variety of ways: as the potential to build on and reshape; as the potential for economic development; as the potential for urban residential extension; or as intrinsic potential and quality for enjoyment. There is a big difference between development potential and intrinsic quality. For many, the fringe has potential only as far is it can be re-used and redeveloped (and essentially made 'urban'); for others, what is there already should be celebrated. Our natural inclination is to take a middle path: we will attempt to show – in Part Two – that the fringe has intrinsic quality and development potential, and then discuss, in Part Three, how this potential might be realized. But conventional planning is likely to have an urbanizing effect on the fringe. While some writers have sought to celebrate what the fringe already has in terms of biodiversity and educational potential (see Shoard, 2002); others would like to see its potential 'repackaged' for a more demanding consumer. The big risk that planning poses in the fringe is that a wild and diverse landscape will be transformed into something very sanitized: an urban theme park of mowed grass, science parks, water features, bespoke farm shops, and designer green-ways traversed by middle-class cyclists with children in tow. Whether the fringe needs sanitizing – for mass consumption – is a question that we return to throughout this book.

Conclusions

The rural-urban fringe is unique, in terms of its mix of land uses, the characteristics of its landscapes and the way it is frequently represented. It is not merely an extension of town into country, or a transitional aberration delaying the onset of real countryside: it is that land lying between urban areas and countryside that has its own separate and frequently unique characteristics. Some of these characteristics have been summarized in this chapter, but the remainder of this book will refine our initial view of the rural-urban fringe.

The book has a number of aims, and a number of underlying themes. At a practical level, our intention is to more clearly understand what the fringe is, what pressures it faces, and how planning might respond to the various challenges that exist. At another level, we will demonstrate how the fringe has been made and how it is perceived and represented. In examining what the fringe is, and how it might be managed, we apply – albeit loosely – the concept of landscape multifunctionality. But a key, recurring, theme for this book is the legitimate purpose of planning at the edge: to realize development potential and orchestrate change, or to promote the qualities of the fringe in a perhaps more sensitive way.

2
The making of the English rural-urban fringe

Introduction

The rural-urban fringe is becoming a focus for planners and policy makers, who have suddenly turned their attention to this hitherto neglected landscape. The green belt – the jewel in the crown of UK planning policy since World War II – may soon be subject to review; housing pressures, particularly in the south of England, are leading many to question the logic of 'protecting' land that often appears derelict and in need of improvement. But others continue to argue that the green belt, and the fringe more generally, needs some form of protection, not because of any intrinsic value, but because building at the edge is likely to stall any hope of an urban renaissance in the inner cities. But a new debate concerning the fringe is also taking root, based on the assertion that planning at the edge – including green belt policy – has never been *for* the edge. Rather the fringe has been subject to a deliberate policy of neglect: this policy has sought to create a buffer between town and country with developmental objectives on the urban side of that buffer and environmental ones within the countryside beyond. But it has never sought or aspired to do anything *inside* the fringe. Today, there is growing consensus that the fringe is an area of untapped potential and opportunity, where unruly dynamism might be harnessed and a multifunctional landscape created that achieves a range of economic, social and environmental goals. But this is not our concern in this chapter.

Rather, we are concerned with the neglect and not the power of planning to do something different at the edges of towns and cities. There is a limited literature devoted specifically to the English rural-urban fringe: a few authors have pointed to the fringe's unique characteristics (notably Shoard, 2000, 2002), but most view it as a transitional landscape that marks a blurring of the boundary between town and country. While Shoard (2000) sees it as an 'edge land of promise', others have focused on how the fringe might simply be turned into something else, perhaps through urban extension, the creation of more regimented park land, or an expansion of urban agriculture. In this chapter we have two key aims: the main aim is to argue that neglect of the fringe is, paradoxically, the key process 'making' this landscape (it follows that planning a hitherto unplanned landscape will radically alter the direction of its future development); a second aim is to join Shoard in arguing that this is a unique landscape with very special characteristics, and not merely a grey area between the urban and the rural.

Making the fringe

The title of this chapter deliberately parallels Hoskins' *The Making of the English Landscape* (1955) and several other books on landscape, which include the word 'making' in the title. Roberts' *The Making of the English Village* (1987), Rowe's *Making a Middle Landscape* (1991), Whitehand's *The Making of the Urban Landscape* (1992) and Trinder's *The Making of the Industrial Landscape* (1997) have all followed the path set out by Hoskins, but they do not share a particular theory about how landscapes are actually 'made' or produced. What they have in common is the basic underlying assumption that landscapes are somehow 'made' and, in different ways, they illustrate how economic, social, political and demographic changes have impacted on and shaped the landscape. We aim to follow this loose tradition, looking at how the rural-urban fringe has been made. However, a key departure from this past literature is that there has been far less deliberative action in the fringe, which has been made more by fortune than by design.

Although Hoskins' book in some ways pioneered the notion that landscapes are 'made', it also created powerful oppositions between urban and rural and seemed to suggest that the countryside was 'immemorial' in contrast to other landscapes (Hoskins, 1955: 231). However, it is important to remember that the countryside is also man-made and subjected to changes that stem from global and local interactions: 'even in Victorian times, fundamental changes in the British countryside were only explicable with reference to wheat imports from the New World, and the dereliction of lead mining landscapes in the Pennines at the close of the nineteenth century was largely caused by developments in the production of lead/silver in Spain' (Muir, 2003: 384). Hoskins was of course aware that the countryside is 'made', but he seemed to distinguish between the dramatic change of his own time and the slower landscape change of previous eras. As Matless has put it: 'No matter that all the landscapes discussed by Hoskins – deserted medieval villages, parliamentary enclosures, country parks – are the product of historical change, in this landscape sensibility modern change becomes somehow different in its erasure of meaning, rubbing out the historic document of landscape' (Matless, 1998: 276).

Hoskins had a clear dislike for the twentieth century, and particularly for the post-war world. In his view, landscapes had been made through gradual social and economic processes. The landscape was layered and told a clear story stretching from the earliest periods of human habitation. But in the same way that landscapes could be made, they could also be unmade, with meaning erased through insensitive planning and development. For Hoskins, much of the meaning bestowed on the landscape through a thousand years of history was obliterated after 1945. This was a sentimental, quintessentially English view of the landscape and of the processes that make villages, medieval towns, and the rolling pastoral patchwork quilt landscapes of middle England. World War II was a turning point: everything that went before was positive, contributing to the safe notion of Englishness: everything after brought rapid change and not for the better. This world view was also reflected in Orwell's 1939 novel *Coming Up for Air*, which sought to contrast the slow change of Britain between the wars with the political and economic upheavals that the author expected might follow a conflict in Europe. It is clear that the notion of landscapes being made, in the sense intended by Hoskins, was heavily value-laden: an analysis of change which also rejected that change when it began to offend what was essentially a conservative view of the English landscape. Implicit in Hoskins' work is a rejection of the political and economic undercurrents shaping the landscape, of the social changes afoot in post-war Britain, of urbanity and of modernism.

But all these processes were integral to making the fringe. Although, as Thomas (1990) points out, the fringe is as old as civilization itself, what we might consider the 'modern fringe' is really a product of those post-war processes deplored by Hoskins. For instance, although the decentralization of light industry, warehousing, storage and retail might have its roots in the early twentieth century, movement of such activities began to accelerate with the construction of arterial and orbital motorways after the war: a process that had already been noted in the United States by Wehrwein in 1942. Decentralization had not been a particularly slow process before the war and many industries, previously tied to central locations, had been encouraged by railway development to exploit cheap land in suburban locations. Again, in the US, Wehrwein (1942: 222) has attributed an acceleration of fringe development to freeway development, to the opening up of cheaper land, to lower taxes on out-of-town sites and, critically, less stringent land-use controls away from inner city areas. In the United States, these were the early drivers of fringe development. But they also have their counterparts in the United Kingdom. The creation of a comprehensive network of motorways between the late 1950s and the early 1970s provided a skeletal framework, shaping subsequent fringe development, playing a similar role to railways in the inner city a century earlier (Figure 2.1). Then, railway terminals, stations, warehouses, depots and goods yards contributed to traffic congestion and the rail tracks themselves fragmented the city and limited access between districts and neighbourhoods. As Trinder (1997: 228) has pointed out: 'railway companies owned between eight and ten per cent of the land in central areas, and influenced the character of up to twenty per cent'.

Development associated with railways, such as stations, hotels and warehouses, tended to be located within the city. In contrast, development associated with motorways (motels, service stations, distribution depots, out-of-town-shopping centres, business parks and science parks) are often located in the fringe. The contemporary rural-urban fringe is in that sense as shaped and fragmented by motoring as the nineteenth-century city was by rail.

30 *The making of the rural-urban fringe*

2.1 Motorways truncating the landscape, M1, north of Milton Keynes (Countryside Agency, David Woodfall).

The rise of the aviation industry and the subsequent expansion of airports is the latest addition to this landscape, which has not only been transformed by transport infrastructure, but is also almost entirely experienced at speed from inside a vehicle of some kind (Kamvasinou, 2003: 182). Even railways are today having a greater impact on the rural-urban fringe with the advent of parkway stations designed to vent some of the pressure on central areas generated by railway development. Those commuters living in the suburbs of smaller towns no longer have to drive to a central station to catch the train to London, but can head to a nearby parkway, find ample parking space, enjoy the facilities of a new station, and continue their commute by train. The expansion of London's outer ring from the 1950s onwards provided the impetus to expand the network of arterial roads truncating London's fringe. Hall (1989) points out that the city witnessed a steady decline in population in the 1950s. But at the same time, London's outer area, defined as a ring with an average diameter of 45 or 50 miles from the centre, experienced massive expansion, growing by nearly 1 million people, in only 10 years. This pattern of change is repeated around other major centres: population decentralization has fuelled the growth of suburban and fringe road networks. Modern transport infrastructure has undoubtedly played a significant role in making the rural-urban fringe, not only shaping its physical appearance but also shaping experiences of it. As Kamvasinou (2003: 182) has pointed out, it is experienced at speed, suggesting a transitional landscape, and a dynamic space that people move through but tend to ignore. It is the antithesis of Hoskins' slow, rambling landscape where nothing much changes and everything – churches, hedgerows, ancient castles, undulating hills, pastures and meadows – is held in high regard.

The first conclusion we can reach is that the modern fringe has been made by the motor-car, by the need for better roads, and by the processes of decentralization and land fragmentation associated with post-war road building. Increased private vehicle ownership and improved road and railway communications around metropolitan cores, have encouraged extended decentralization within the largest established major urban systems, and consequently reshaped the appearance of the rural-urban fringe. According to Cross (1990: 4):

> initially, the areas with the most rapid inflows were suburban but more recently growth spread over a broader area. First growth has reached into adjacent areas of the region, through an expansion of the limits of the commuter fields, and then from larger metropolitan areas into smaller daily urban systems, principally around conurbations.

But the roads are there, fundamentally, to serve the city or satellite towns: orbital roads afford faster access to different parts and arterial routes allow rapid access or exit. The motorways serve the city and not the fringe: hence the development that has sprung up in what some have termed 'interstitial' landscapes, associated with road building, is there more by fortune than design. In many respects, this is an accidental landscape which, by accident, has taken on a life of its own during the last half century, 'made' by a desire to serve the city with road building and protect the countryside from sprawl. These are perhaps the most obvious processes making the fringe: but there are others, that are arguably more subtle and it is to these – and to the creation of a unique fringe landscape – that we now turn.

A unique landscape

The term 'urban fringe' is usually attributed to T. L. Smith, who first used it in relation to 'the built-up area just outside the corporate limits of the city', in a 1937 study of population changes in Louisiana. One year earlier Herbert Louis had coined the term 'fringe belt' (in the German original *stadtrandzone*) in a study of Berlin (Thomas, 1990: 132–133). A 'fringe belt' is an area where 'peripheral land uses, when later encompassed by the outward growth of the built-up area, tend to survive as a belt, sometimes continuous, separating older from younger development' (Whitehand, 1967: 223). Although these terms date back to the 1930s, some of the industrial land uses we still associate with the fringe have their origins in the nineteenth century: it was at the time of the industrial revolution that the expansion of settlements into the countryside gained speed. By the time of the industrial revolution, as Picon has pointed out, 'the city stops being in the landscape, as a sort of monumental signature, to become, progressively, in and of itself, landscape' (Picon, 2000: 67). And by now, he continues: 'the city seems to have absorbed the countryside around it; it seems to have made itself master in order to form with the countryside an amalgamation of a new type' (ibid.: 68).

The motorway development discussed briefly in the last section was, in a sense, merely a new phase in the long-standing relationship between town and fringe, feeding the capacity of cities to absorb the countryside in their immediate proximity, blurring the boundary between town and country (Figure 2.2). This idea of an amalgamation between town and country is a recurrent one and the rural-urban fringe is often 'defined as the area of transition between well recognized urban land uses and the area devoted to agriculture' (Wehrwein, 1942: 217). When the architectural journalist, Nairn, attacked what he called 'subtopia' in the 1950s, the emphasis was also on the fringe as a hybrid of country and city:

> subtopia, a mean and middle state, neither town or country, an even spread of abandoned aerodromes and fake rusticity, wire fences,

2.2 Transition from 'rural' to more 'urban' land uses, New Acres, County Durham (Countryside Agency, Mike Williams).

traffic roundabouts, gratuitous notice-boards, car-parks and Things in Fields. It is a morbid condition which spreads both ways from suburbia, out into the country, and back into the devitalized hearts of towns . . . subtopia is the world of universal low-density mess.

(Nairn, 1955: page unnumbered)

However, the idea of the fringe as a hybrid, an amalgamation or a zone of transition with urban uses gradually giving way to rural uses and vice versa, somehow fails to recognize the uniqueness of the urban fringe, or the particular assemblage of processes that make this landscape. The transition model seems to suggest that offices and factories (ostensibly 'urban uses') show a gradual transition from a high concentration in urban centres to a lower concentration in the fringe and a very low density in rural areas. However, farm diversification in recent years has seen conversion of farm buildings to business and industrial uses and fringe areas in some regions contain more industrial, office and retail use than town centres. As Punter has pointed out, Nairn's attack on 'subtopia' seemed to represent a strong attack on the visual products of planning:

> though the author quickly discovered that many of his targets were, in fact, exempt from planning control. The unwitting agents of subtopia were correctly identified as local authority departments (highway engineers, public works departments etc.), statutory undertakers (Electricity Generating Boards, Forestry Commission, etc.) and Government Ministries (Ministry of Defence, Ministry of Transport) [for road schemes and military airfields], and the fragmentation of control was well illustrated.
>
> (Punter, 1985: 15)

This fragmentation of control has resulted in a fragmented landscape, and many fringe areas seem uncoordinated and unplanned. In the United States, Carruthers (2003: 478) has argued that 'political fragmentation affects development patterns by dividing land use authority among numerous individual jurisdictions'. Freidberger (2000: 507) has argued that developers have thrived in the 'boom-like atmosphere of the rural fringe where regulations for land control were minimal'. A lack of coordination, resulting in unplanned, feverish development in some instances means that the rural-urban fringe is not simply an amalgamation or hybrid of town and country, but represents a completely different spatial organization of land uses.

Spatial (dis)organization as the product of uncoordinated planning is an important driver making the post-war fringe. The splintered politics of the urban edge noted by Punter (1985) has resulted in responsibility being shunted and shunned. Many statutory undertakers enjoy a greater range of permitted development rights without recourse to local authorities; the same is true of the Ministry of Defence and other Ministries with an interest in the fringe. At the same time, the fringe may fall at the edge of local authority responsibility; there may be no local voters and therefore no voices calling for improvements in environmental quality. Local authorities may give up on the fringe, or may find it particularly difficult to manage uses – including Nairn's aerodromes – that they do not really know what to do with (Lober and Gallent, 2005). In some ways, this post-war power vacuum contrasted with the processes making the fringe (or rather, making the suburbs) prior to 1939, in the period before the green belt. For example, Peter Hall and colleagues (1973) observed that at the beginning of the nineteenth century, urban areas remained fairly compact, with city boundaries ending sharply against open countryside (Hall et al., 1973: 76). A lack of cheap mass transport resulted in population densities at a level that the Urban Task Force (1999) would like to return to today. The arrival of such opportunities in the inter-war period – and the lack of any coordinated containment strategy – resulted in a march of housing development along arterial roads or public transport routes and frequently took the form of low-density 'ribbon development' extending well beyond the more continuously built-up area. In 1930, for the first time in Britain, over 100,000 private houses were built in one year without subsidy (Ward, 2004) and to a large extent, up to 1939 mass transportation in most British cities remained largely radial (Hall et al., 1973). Along with a conversion of rural land to urban and particularly residential uses, this aspect of suburban development caused public and political concern. During the 1920s a wide range of preservationist groups started to campaign against development impinging on the countryside (Williams-Ellis, 1975 [1928]). The Campaign for the Protection of Rural England (CPRE) was created in 1926 and the idea of limiting urban growth, giving a sharper edge to the city to allow the preservation of rural landscapes for scenic and recreational enjoyment, began to gain momentum. But herein lies the paradox: inter-war planning resulted in residential sprawl, halted by the green belt after the war, but the green belt has failed to contain fringe development. In a sense, the fringe plays host to two planning extremes: green belt on the one hand, and an uncoordinated, splintered and fragmented, brand of planning on the other.

The original concept of clear physical boundaries between large cities and their surrounding countryside dates back to the late nineteenth century when Howard introduced his proposal for managing growth in London and maximizing accessibility to adjacent green land (Howard, 2003 [1898]). His ideas were further developed by Raymond Unwin in the 1930s and by Patrick Abercrombie in the Greater London Plan of 1944. Abercrombie's plan for the London region provided for a discontinuous park belt – he called it a 'green girdle' – with town and village extensions set deep in the surrounding countryside to receive population displaced from the conurbation. In the post-war period, this approach received statutory blessing and has since been one of the cornerstones of land-use planning. Arguably, the green belt – where it exists – has been bolted onto cities and although incumbent local authorities often fiercely reject residential development within green belts (Elson, 1986), they effectively turn a blind eye to the activities of statutory undertakers developing essential service functions such as electricity sub-stations, sewage treatment facilities and so forth. Post-war planning has become stronger, in many respects, but the needs of cities have ballooned and the fringe has to play host to a range of service functions irrespective of the green belt designation.

We noted earlier that planning at the fringe has not necessarily been for the fringe: we would add now that containment strategies, in particular, have not only ignored the needs of the rural-urban fringe, but have also delivered unintended results. Containment has both diverted attention from the fringe, and also driven change – or created the conditions for change – within the fringe. Containment within settlement boundaries has generated future hope value (that boundaries will change, that the green belt will reform, and that planning will relax) based on the assumption that planning cannot contain growth ad infinitum: something will have to give, at some point. The tension between containment and land speculation has created a fragmented landscape at the rural urban fringe. In the context of the United States, Archer (1973: 367) argues that

> one of the main causes of . . . scattered development is land speculation whereby land is purchased and held for resale at a later date at a higher price. Farmers and developers can also act as speculators by withholding their land from development waiting for higher prices. These landowners hold their land out of the current market so that land and building developers have to bypass it and homebuyers have to travel further afield to purchase new housing.

In the UK, this pattern of speculation will not have an impact on the landscape until planning restrictions are lifted and land is earmarked for development. It will only affect the landscape where boundaries are pushed outwards in successive development plans and this does not apply to land that remains under green belt restriction. But even within green belt land, a tension between restriction and the desire to develop plays its part in the making of the fringe. For instance, the untidiness of land on the urban edge often signals the desire of landowners to secure development rights. Land is purposely left open to illegal fly-tipping in the hope that it becomes so degraded and such an eyesore – in the view of nearby residents – that the local planning authority will start to question the value of preventing development and might eventually request a review of local green belt boundaries. Owners may be encouraged to go down such a road by companies publicizing the scale of housing pressure, especially in southern England, and prophesying that as pressure builds, land – including green belt land – will need to be released. JP Land Sales, for examples, notes that 'there is a finite supply of land in the UK. As more people are realising the investment potential of good quality, strategically located land, the supply is diminishing. Yet as building pressure on the land around our towns and cities continues to grow and the land values rise, demand continues to soar' (http://www.jplandsales.com/land-investment.htm; accessed 11 May 2005). Many speculators – and planners – no longer believe that the green belt is sacrosanct. Pressure at the edges of towns and cities in part explains the current surge of interest in the rural-urban fringe, and anticipation of change is becoming a force, in its own right, shaping the landscape of the rural-urban fringe.

This is very much here and now, and we return to the issue of the green belt – and its possible futures – towards the end of this book. It is also the case that pressures at the urban edge contribute to a fringe landscape only while those pressures are resisted. Once the floodgates open, and residential development extends beyond the built-up area, the fringe ceases to be fringe, and takes on suburban characteristics. But while there is continued resistance to this development, the fringe remains a 'landscape without community'. It is to this major characteristic of the fringe that we now turn, while retaining our focus on the processes making the fringe.

Landscapes without community

The fringe is largely a landscape without community, but which people experience in a range of different ways. For some, the experience might be positive, crossing the urban edge to tend an allotment. For others, experience is entirely functional, and probably ranges from neutral to negative: it is a drop-off point for a parkway station, a retail park, an office complex or an airport. There may not be any specific purpose for accessing the fringe other than to pass through it on the way to somewhere more 'meaningful'. A landscape without community is arguably a landscape without a voice, anonymous, functional and cold. The political goals of the New Right in the 1980s contributed to the anonymity of the landscape: a rampant free-market ideology saw the fringe opened up to commerce, especially out-of-town retail. For Sinclair, the M25 – opened by Margaret Thatcher in 1986 – epitomizes the soullessness of the fringe:

> The Leatherhead stretch, on this July evening, is leisured, a mini-autobahn, a military highway of the kind Margaret Thatcher fantasised when she cut the ribbon. The principal difference, so far as I can see, between the Thatcherite Vision of the Eighties and National Socialism in the Germany of the Thirties is that Thatcher couldn't make the trains run on time. The M25 never was an invasion route down which the master race could roll, just a three-hour fairground ride with dull views.
> (Sinclair, 2002: 270–271)

The idea to build an orbital road for London certainly pre-dated Thatcher. The Abercrombie Plan of 1945 (Abercrombie, 1945: 78) proposed a series of five individual roads around the capital and over time successive governments cut back this original vision until construction began in 1975 under Harold Wilson's Labour government (Hall, 1994). Eleven years later, the vision was reality. We have already noted the fragmenting role of road-building in the fringe and the daily misery-go-round between the suburbs, parkway stations, retail parks and so forth (Figure 2.3). The intention here is not to cover this ground again, but to reflect on the way a lack of coherent community is itself a driver of change within the fringe. There is little literature dedicated to this

Landscapes without community 37

2.3 A landscape fragmented by roads – A69 bypass, Haltwhistle, Northumberland (Countryside Agency, Charlie Hedley).

issue, but we draw inspiration from both the M25 (perhaps the ultimate in uneighbourly but essential uses) and from the wave of NIMBY pressure that has swept across the English home counties since the end of World War II.

The draft version of the South East Plan (South East England Regional Assembly, 2005) sets regional development priorities within the London Fringe, the Kent Thames Gateway, land abutting the Western Corridor of the M4, in an area abutting Gatwick Airport, and in several other strategic locations. These listed areas have a higher share of previously developed land (PDL) and fall within the shadow of existing urban areas. They are all 'fringes' if we apply the label rather loosely: and they are all characterized by low-density non-residential development. None are areas of strong NIMBY sentiment, and hence all are politically acceptable growth poles. The parallel with what can happen in fringes more generally – i.e. development directed to the most politically accepted location – is not entirely fair: these growth areas will be called upon to absorb planned housing growth, which is not 'uneighbourly' in the strictest sense, but which nevertheless provokes a great deal of middle-class opposition. The question we are leading up to is this: why are fringes called upon to host everything that cities – and the traditional countryside – does not want? The answer is simple: there is often no *in situ* opposition to this development.

This is of course an over-simplification. When development is in sight of residential areas, it can be fiercely opposed and hence there has been a sustained conflict over the expansion of Heathrow Airport. But generally, people accept the situating of essential services in the fringe (sewage treatment works, electrical sub-stations, etc.) that would spark vocal opposition in a residential area. Obviously, land-use planning is intended to prevent the mixing of such uses, but there are instances when proposals are brought forward for new waste incinerators considered too close to residential areas. These are fiercely opposed. Parallels can be drawn, again, with the encroachment of railways into nineteenth-century cities. Trinder (1997: 230) has argued that the expansion of the railways into urban cores in the second half of the nineteenth century 'usually affected working class districts, where opposition was likely to be ill-organized'. In much the same way, roads and airports (as well as other less favoured land uses such as nuclear power stations, sewage works, waste processing centres, incinerators and – more recently – immigration reception centres) have been placed in the rural-urban fringe where opposition is likely to be weak or non-existent. Since the fringe – almost by definition – is a place where nobody lives, there is no natural community or electorate that will protest or bring politicians to account for decisions that have affected these particular areas. This has affected not only the type of land uses that occur in the fringe, but also the aesthetic *quality* of developments, whose functions are less controversial. Shopping malls do not have to be of poor architectural quality, but if they are located in the fringe, aesthetic considerations tend to be taken less seriously. It is, for example, unlikely that a temple to kitsch and gaudiness such as the Trafford Centre outside Manchester would have been granted planning permission in or near an existing town centre, or close to a vociferous middle-class community. Historic environments are protected through legislative measures, and local electors give politicians trying to push through unpopular development a hard time.

This raises another important point, and another crucial question. Fringe landscapes are the products of political acceptability and expediency. This is clear from the preceding discussion and this is a consequence of fringes being landscapes without traditional communities. But what is political acceptability? It has a number of levels: it starts off as an aspiration guided by prevailing wisdom. Today this wisdom says that places should have the right jobs–homes balance, that development should have the lightest possible environmental footprint, that private car use should give ground to public transport, and that development should be focused on previously developed land, and so forth. But political acceptability at the local level shifts from being a concept towards being a practicality. It is less of an aspiration and really a measure of local response. The fringe is politically acceptable not because all the goals of government can be achieved at the fringe, but because NIMBY elements are persuaded by the benefits of sweeping development away from residential areas, and the fringes themselves have no opposing voice. This is an observation on process but not a judgement on outcome. Often, the fringe is the best location for certain uses given the characteristics of these uses. But one unfortunate outcome is the lack of concern for good design, noted in the case of Manchester's Trafford Centre. The fringe not only lacks community, but also lacks a champion. This has resulted in a proliferation of low-quality development: bland white warehouses serving a multitude of retail and distribution functions, dumped into a political vacuum at the fringe.

Landscapes without politics

We argue later in this chapter that economic restructuring has created a transitional landscape of redundancy at the rural-urban fringe: it has also – in more recent years – created a landscape of consumption. But we use the rise of out-of-town shopping – an important aspect of consumption – at the fringe to illustrate another point, leading on from the last section: that political acceptability is a function of community acceptability. Without community, there is no issue of acceptability and therefore a political vacuum. This is a rather simplistic view point, but we have already suggested that planning at the fringe is not *for* the fringe, and it only takes a short jump in logic to suggest that the fringe is out of sight and out of mind in a number of respects. We are not arguing that the fringe is apolitical, or that it is not a product of political intervention, but simply that it has become an acceptable dumping ground where decisions provoke little or no political backlash. This is illustrated by out-of-town shopping. The explosive expansion of out-of-town shopping developments from the early 1980s onwards provoked a defiant defence of town centres, and an eventual reversal in policy: it never, however, resulted in concerns over the level or quality of development in the rural-urban fringe. Regional shopping centres, retail parks and factory outlets are now an integral part of the landscape. An increase in car ownership and lack of space and high rents in many town centres may be central factors that have triggered these developments, but as Guy (1998) has pointed out, land-use planning and the changing political climate have also played an important role:

> Land use planning policy in the UK has been an important influence on the pace and nature of off-centre retail development. Initially, a very cautious approach, protective of town centres, restricted off centre non-food retailing to sales of DIY and other 'bulky' goods. In the early 1980s, many local authorities began routinely to allow sales of electrical goods from off-centre locations: in retrospect, this change in policy, which was hardly discussed at all in the practice literature at the time, was crucial. It raised the level of trading density obtainable from retail warehouses, thus encouraging growth in rent demands and capital values. It also set a precedent for allowing 'High Street' retailing in off-centre locations.
>
> (Guy, 1998: 306).

Pressure to promote off-centre retail led to a relaxation of planning which 'gave developers a window of opportunity to promote new retail formats in new locations' (Lockwood, 2001: 13), which were usually 'car-centric', purpose-built to cater for car-borne shoppers. Such developments are frequently rejected today on the grounds that they would have an adverse trading impact on nearby town centres (Guy, 1998: 295). There is of course a line of defence for out-of-town retail: pressure may be vented from congested centres; and factory outlet shops can become tourist destinations, providing a boost to the local economy (Jones and Vignali, 1994: 10). There are both positive drivers steering retail to fringe locations, and a certain de facto reality of fringe planning. McArthur Glen, one of the largest factory outlet developers in the US, 'stresses the importance of high visibility locations, ideally with good motorway junction access' (ibid.: 9). For similar reasons, the ideal location for regional shopping centres is in the fringe: 'the very nature and *raison d'être* of regional shopping centres explicitly demands access to large catchment populations and thereby *de facto* assumes considerable urban linkage' (Lowe, 2000: 270–271).

There are numerous centres in fringe locations that follow these basic rules. Brent Cross in northwest London, at the southern end of the M1 motorway on the North Circular Road, was Britain's first purpose-built 'out of town' shopping centre. It opened in 1976 and comprised 800,000 square feet of retail space (Jackson, 1999: 30). In the 1980s four similar developments followed: Metrocentre in Gateshead, Meadowhall in Sheffield, Merry Hill near Dudley in the West Midlands and Thurrock Lakeside east of London. Meadowhall is a particularly interesting case, 'built directly on the site of a major local steelworks [at Hadfield], which in 1980 had been the site of the biggest mass picket in the steel strike of that year, but which closed three years later' (Evans et al., 2000: 158). Along with new developments in Bristol and Leeds, the 1990s saw the opening of Britain's two most spectacular shopping centres to date: the Trafford Centre in Manchester and Bluewater near Dartford in north-west Kent.

Returning to the point made earlier, these centres exploit a planning regime at the rural urban fringe that has, in the past, turned a blind eye to car dependency, and that continues to display scarce regard for design quality. A dearth of policy planning in the 1980s and 1990s has been bridged by a tougher regime today, expressed through 'PPS 6: Planning for Town Centres' (ODPM, 2004a) which establishes a sequential test for retail, calling on planning authorities to adopt an essentially protectionist stance towards town centres, utilizing central sites for retail before considering the release of land at motorway junctions or other similar fringe locations. Clearly, there is an emergent fringe politics here, but reflecting a prioritization of town centre rather than a concern for the fringe per se. The fringe remains a safe haven for activities that might be frowned upon elsewhere. Perhaps labelling the fringe a landscape without politics is a little too strong, but it has certainly been an area where rules have been bent, regulations circumvented and policies paid considerably less lip-service. There is some evidence that this is changing today, partly because of a resurgent concern for the fringe, as planners and commentators start to ponder a world without green belts, or a world in which local planning becomes more involved in shaping the edge through the use of strategic gap and green wedge policies (see Chapter 8). There is a growing politicization of development activity at the rural-urban fringe, but this is born not from local interest, but from the current (quiet) stampede to shape the future policy agenda at the edge, filling the vacuum that has been characteristic of the recent period of containment.

Transitional landscapes of production, redundancy and consumption

No analysis of the making of landscape would be complete without some consideration of economic process. This issue has already been given passing attention earlier in this chapter, which has considered urban growth pressures and consequent decentralization, edge pressure on the speculative land market, and the movement of retail space. Myriad economic processes have altered the appearance of the English landscape during the past fifty years:

perhaps the most obvious relate to the changing fortunes and practices of agriculture. The move to subsidy-based farming after 1947 (the year of the Agriculture Act) and the emergence of capital and revenue subsidy, resulted in a rationalization of the sector, a concentration of activity into fewer hands and an intensification of farming practice that, it has been claimed, has wrought irreparable damage to land and wildlife (Shoard, 1980). At the same time, there has been a concentration of manufacturing activity in smaller towns, and a disappearance of craft industries from the countryside. Added to these manifestations of economic change in rural areas have been the broader processes – including economic globalization – which saw the profitability of British manufacturing decline from 1945 onwards, and a progressive move to a tertiary, service-based national economy. This has resulted in a reconfiguration of economic space, causing widespread physical decline in many of the heavy industrial heartlands, decimating mining areas, and most recently, bringing a hundred years of volume motor manufacturing to an end. As a consequence, urban areas, rural areas, and the rural-urban fringe look very different today than they did fifty years ago (Figure 2.4).

During the inter-war period there were roughly as many people employed in the agricultural sector as in the mid-nineteenth century. But after 1945 the number of farmers declined rapidly as a result of 'the amalgamation of farm holdings and the growing size of farm businesses' (Cherry and Rogers, 1996: 113). Shoard has argued that the 'executioner' of the English countryside 'is

2.4 Out-of-town shopping, Evesham, Worcestershire (Countryside Agency, Rob Fraser).

not the industrialist or the property speculator, whose activities have touched only the fringes of our countryside. Instead it is the figure traditionally viewed as the custodian of the rural scene: the farmer' (Shoard, 1980: 9). According to Shoard the intensification of agriculture has destroyed hedgerows, heathland, woods, ponds, streams, marshes and meadows and turned the countryside into a 'prairie' (ibid.). The rationalization of farming might have contributed to a monotonous, intensively farmed landscape, removed from traditional nineteenth-century images of rustic and pastoral England, but in the rural-urban fringe – where farming is just one amongst many land uses – the physical and aesthetic impact has, arguably, been less radical. Much more radical, however – in terms of making the fringe – have been the economic changes affecting those manufacturing and primary industries associated with urban areas. In the countryside, a failing agricultural sector and a withdrawal from intensive production have been marked by a transition towards growing *consumption* of the countryside. Intensive production has been reined in by growing environmental awareness in the 1980s and 1990s, and by a desire at the European level to avoid unwanted production surpluses. This has resulted in an economic turn towards consumption, with rural areas becoming a focus of tourist activity, farms diversifying into this sector, and a transition to service industry, following in the footsteps of towns and cities. However, the economic changes affecting manufacturing and primary industry have had a very different effect at the fringe, resulting not in a landscape of consumption (at least, not initially), but in a landscape of redundancy.

We suggest – in Chapter 4, focusing on the 'historic fringe' – that the broad processes of economic change, and a decline in Britain's traditional manufacturing base, have created a fringe characterized in many instances by industrial dereliction. Arguably, the tailing off of manufacturing in the twentieth century – and subsequent changes in the occupational class structure – has affected inner cities and towns to a greater extent than the fringe. The fringe has been subject to *physical* change – sometimes on a grand scale – where steel plants, coal mines and quarries have closed. But in cities themselves such changes have also impacted on communities and on the reconfiguration of economic space. In the London Docklands, for example, a manufacturing landscape erased by global shifts has been replaced by high-rise office blocks hosting a rapidly expanding financial sector. The traditional dockland communities have been obliterated, with low-rise council housing and terraces giving way to the latest in high-income waterfront living. In rural areas, economic restructuring created what Clout (1972) referred to as 'encapsulated communities' comprising the remnants of old occupational communities juxtaposed to a new economic class, deriving income from an urban-based service sector rather than agriculture. The same reconfiguration of communities can be observed in the Docklands (Brownill, 1990). Because the fringe has been without communities, it might be argued that the impact of economic change has been less pronounced on the physical fabric; but, alternatively, it could also be argued that the impact has been greater. The fringe was entirely silenced by economic shifts, buildings abandoned, left to rot, creating a potential that only now is being recognized, if not realized. Government responded to industrial decline in the cities with a range of regeneration initiatives: it tried to prop up agriculture with capital and revenue subsidy. But there has been no deliberate response to the physical and economic decline of the fringe. Hence, economic shifts simply created a landscape of redundancy.

But this redundancy has not persisted everywhere. We have already considered the movement of retail to the fringe in the 1980s and 1990s. Arguably, the physical landscape of the fringe has become one characterized and made by consumption. This point returns us, full circle, to where we began this chapter: mobility and the car have made the rural-urban fringe. Nowhere is the physical impact of road infrastructure greater than in the fringe, where six-lane motorways divide the landscape and punctuate it with sprawling junctions, roundabouts, and concrete flyovers. The original thinking behind post-war road-building was to afford greater access to places lacking rail links (Stathis, 1986: 41). In reality, road building has created an insatiable desire for the motor-car and has become a defining icon of the rural-urban fringe. Added to this is the sharp rise in consumer spending – especially since the 1960s – which has fuelled growth in the passenger aviation industry and airports, which have become another iconic feature of the fringe. While earlier airports such as Gatwick, which was completed in 1936, had been located near an existing railway link to London (Voigt, 1996: 47), later airports were often located near motorways, bringing together two key ingredients of fringe landscapes. And finally, the consumer society produces ever more waste, and the fringe has become the natural and preferred location for handling that waste. The fringe is a landscape made by consumption, and scarred by the detritus of that consumption.

But these are really the means and products of consumption. In the contemporary fringe, a complex web of motorways and trunk roads leads to multiplex cinemas, factory outlet stores, garden centres, tenpin bowling alleys, shopping malls, and to Ikea, the ultimate symbol of the flat-packed consumer age. All of these destinations, together with business and science parks, have been controversial from a design point of view, as we have already noted. A feature in *Architectural Review* described business parks as 'any tawdry collection of B1 buildings (those permitted for both office and light industrial use) shoehorned into a desolate roadside site with a landscape consisting of a patch of grass and hornbeam hedge' (Pearman, 1991: 61). Science parks, which emerged in the UK in the early 1970s, were seen as catalysts for regeneration (Westhead and Batstone, 1998: 2198) and have a similar reputation for architectural blandness and uniformity (Figure 2.5). Castells and Hall have suggested that the image of the science park represents our current information technology economy. If 'the coal mine and its neighbouring iron foundry, belching forth black smoke into the sky' was the image of the nineteenth-century industrial economy, the corresponding image for our contemporary economy 'consists of a series of low, discreet buildings, usually displaying a certain air of quiet good taste, and set amidst impeccable landscaping in that standard real-estate cliché, a campus-like atmosphere. Scenes like these are now legion on the periphery of virtually every dynamic

2.5 Science park in its landscaped setting, Evesham, Worcestershire (Countryside Agency, Rob Fraser).

urban area in the world' (Castells and Hall, 1994: 476). The anonymity of these places stems from the fact that they look the same all over the world and, in that sense at least, they correspond to Augé's notion of 'non-places' which 'do not integrate the earlier places: instead these are listed, classified, promoted to the status of "places of memory", and assigned to a circumscribed and specific position' (Augé, 1995: 78).

Ultimately, this is the landscape of the rural-urban fringe, the product of mobility, of rapid change, of splintered politics, of containment and land speculation, uncertainty, a lack of community, of political expediency, and of economic shift, redundancy and consumption.

Conclusions

In a very approximate sense, we can claim that the landscape of the rural-urban fringe has been 'made' by an array of social, economic and political processes. The same is true of all landscapes. There has been little past concern for the making of the fringe, which has been viewed as a transitional landscape subject to rural and urban processes, and 'different' only in so far as it mixes these processes together to create a hybrid space. In this chapter, we have tried to show that unique processes have impacted on the fringe, and these have been juxtaposed with more general (e.g. economic) processes that have nevertheless affected the fringe in a very particular way.

Conclusions 45

First and foremost, mobility, the motor-car and road-building have made the fringe in the sense of giving it that 'transitional' and rapid feel so commonly associated with the urban edge. This has created a dynamic landscape that sits in marked contrast with the countryside beyond and the city within, which can seem turgid and immovable in comparison. The splintered politics of the fringe has frequently resulted in uncoordinated planning and action, with different agencies doing different things with little in the way of joined-up thinking. For planning authorities, a landscape without community can be ignored: it may become politically acceptable and expedient to push less desirable land uses and activities to the fringe. This suggests an apolitical landscape where planning and policy affect the fringe by virtue of pursuing objectives elsewhere (retail may become less common in the fringe because of a desire to protect High Streets). And finally, the fringe is a landscape made by economic change that resulted first in redundancy, and then, in consumption. Consumption is the product of opportunity, created by the roads that sprang up in response to demographic decentralization in the post-war period (Hall, 1989). Economic change fuelling demographic shift, requiring road building, providing consumption opportunities following a period of redundancy, 'made' the fringe in the loose sense employed by Hoskins (1955), Roberts (1987), Rowe (1991), Whitehand (1992) and Trinder (1997).

Many of these processes are reconsidered later in this book as we examine the nature of the rural-urban landscape and the functioning of landscapes. Key issues are also picked up again in Part Three where attention turns to past and future planning at the edge.

Part Two
Multiple fringes

3
A historic fringe

Introduction

All landscapes possess historical association: this may be expressed overtly or more subtly. People are drawn to historical cityscapes, to stately homes, medieval castles or to ancient world heritage sites. Many such places experience intense visitor pressure and are key features of national or even global place-marketing strategies. But how many people are drawn to the urban edge? There are of course key sites that can be found on the edge of London: Epping Forest with its Elizabethan hunting lodge or Hampton Court which might be described as a historic jewel on the capital's periphery. But these are not the subtle landscapes of the rural-urban fringe, where history does not announce itself with a fanfare, but is often tangled within a patchwork of redevelopment or has itself taken on a new role. In this chapter, we intend to show that the fringe is a historic landscape and often the product of the ebb and flow of industrial, agricultural or military development at the edges of many towns and cities. We will argue that the fringe is host to a rich and diverse archaeological and historical legacy that is not always fully appreciated, or valued.

Many of the features of the historic environment that are found in the fringe today may be the products of its earlier rural existence, its previous history as a dispersed settlement in prehistoric times, or as a place for industrial and service activities related to the growth of the urban centre that it now envelops and serves. Some of these features may have an anticipated re-use value: we noted in Chapter 1 that old airfield hangars have often been re-used today for light manufacturing or for industrial storage. Former cotton mills in the north of England – especially those surrounding the former Lancashire cotton towns – may come to serve a similar purpose. Similarly, canals penetrating into the fringe have in many cases been rejuvenated by local volunteer groups and now play a leisure rather than a commercial role within many local and regional economies. The Grand Union Canal running between the West Midlands and London is perhaps a classic example, skirting many south Midland towns and providing the incentive to keep numerous peripheral pubs and parks in business. But in contrast, many of the canals of Birmingham's industrial fringe have fallen into disrepair and, arguably, a city that boasts more canals (by length) than Venice has not made best use of this resource (except in the more central locations where, for example, canal-side redevelopment has played a major part in the Broad Street regeneration).

Whereas many central or more rural historic buildings have been conserved for conservation's sake, the historic features of the fringe – less visited and perhaps less visible – are frequently valued only when they can be re-used. Apart from a scattering of preserved Battle of Britain hangars, many wartime or cold war military sites have survived only because of their re-use potential. They are rarely conserved as visitor attractions, perhaps because they are off the beaten track, or because we simply do not believe that such places can be marketed. There are of course many high-profile exceptions, where industrial heritage – including the dark, satanic mills of the north of England – does draw visitors. In South Wales, people flock to the former 'Big Pit' colliery (closed in 1980) near Blaenafon to experience what many South Wales miners spent a lifetime hoping to escape: the deep coal pits of the early twentieth century. The great danger at the fringe, and more generally, is that we protect only those things that we can fix a price tag to. Canals are protected because developers believe that a canal-side location places a valuable premium on new development; and the shells of buildings dating from the industrial revolution or from World War II are preserved only if re-use is commercially viable (and demolished if it is not, so long as a listing does not get in the way). Conservation has a strong 'here and now' slant; judgements are made today on the future of buildings and features that may have greater value tomorrow. Using this argument, all historical features have potential value that is more fundamental – and greater – than re-use value. In this chapter, we aim to show that the fringe – the high and the low water mark of urban development – is rich in a heritage that should be conserved (Figure 3.1). But it should not be conserved to achieve any single goal: rather, history and heritage should be acknowledged in all planning and management strategies for the fringe: sometimes this may *lead to* re-use, sometimes to the creation of new visitor resources, and sometimes to an educational resource that imparts a greater understanding of the historical development of place.

Again, we would stress that cities have their great buildings and cathedrals and rural areas their castles and houses, but what is the historical architectural legacy of the rural-urban fringe? Although some traditionally 'grand features' such as country mansions and castles do occur in the fringe, the built environment on the edges of towns and cities predominantly consists of functional structures associated with an industrial or military past. The aim of this chapter is to elucidate the historical and archaeological richness of the fringe – a richness we believe is frequently misunderstood and undervalued – and, more specifically, to provide examples of how this historical landscape adds to multi-functional potential and can be re-used in a variety of different ways. We have already mentioned Colliers Moss near St Helens in Chapter 1, a contemporary fringe area that is the product of its ex-industrial legacy, where former colliery

Introduction 51

3.1 The fringe is rich in industrial archaeology – Devon (Countryside Agency, Chris Chapman).

buildings have been retained and re-used, where machinery has been integrated into environmental projects, and where historic, economic and ecological fringes have been brought together in a way that seems to accord with the notion of multifunctionality. In this chapter we expand on this and introduce other examples of how the historical legacy of the fringe can be highlighted and incorporated into regeneration strategies for the fringe. We also aim to illustrate that the fringe has a strength of historical association that cannot be ignored in any planning or management at the edge.

More particularly, this chapter traces the roots of the historic fringe from the pre-modern period before turning to look at the nineteenth and twentieth century industrial and military legacy which, arguably, provides important and often commercially viable opportunities for contemporary re-use.

Pre-modern archaeology

The fringe is most obviously a product of post-war process, of urban expansion, road building, the rapid development of infrastructure to serve growing urban populations and so forth. Any urban area that is now larger than it was in the past is likely to contain relics of former fringe uses. York in Roman times, for example, is one place that has been extensively studied. There, archaeologists have found evidence of four uses just outside the boundary of the Roman fortress and the civilian town – pottery and tile manufacture in kilns (which would have been noisy and smelly), military training at practice camps, burial and rubbish deposition. However, these remains now lie under the later town of York and do not tend to be located in its modern fringe (though they were fringe features of a town that occupied a different footprint). This sort of picture is likely to be repeated in many locations; excavations at Chelmsford, for example, have revealed a very similar situation (Shoard, 2002). Pre-Roman fringe features are, of course, much rarer. This is because very few modern settlements predate the Roman occupation. Stone, Bronze and Iron Age settlements have now been wiped from existence, have been completely developed over, or have left remnants many miles from existing towns. However, some have left legacies close to modern towns and cities and there is evidence of pre-Roman settlement within some fringes. In Wales, the town of Aberystwyth stands to the north of a much older Iron Age fort at Pendinas and there is evidence of former occupation on the banks of the River Rheidol where Iron Age settlers discarded mussel shells and other evidence of their daily lives. But these are not fringe features, though they happen to be located in some contemporary fringes in a few instances. In the case of Aberystwyth, the settlement shifted location from the banks of the Rheidol to the south, to the banks of the Ystwyth just to the north (the two rivers enter the Irish Sea at the same point). This relatively short jump in location has left a historical legacy in the fringe. In instances where the modern settlement shares location with what has gone before, urban expansion has often obliterated what evidence there was of former occupation. But this is not always the case: in some instances, previously dispersed settlements may have contracted into a smaller nucleus over time. Some English towns stand on the sites of dispersed prehistoric towns that eventually became nucleated only to expand again in the nineteenth and twentieth centuries. Chichester on the English south coast, and Colchester in Essex both display evidence of this type of morphology. Before Colchester became an important Roman fortress, it was a late Iron Age royal centre spread out over a larger area. This means that the present-day rural-urban fringe of Colchester takes in land settled in prehistoric times but subsequently abandoned when the settlement became more concentrated. The rural-urban fringe of such towns can therefore contain a great deal of archaeological interest.

Often what we find at the fringe predates the existing settlement and is evidence of a history unassociated with the town or city that developed in a later period: indeed, the archaeological importance of the fringe may relate to a time when the area was entirely rural. The planning system has not always responded sensitively to such landscapes. In the early 1970s, Springfield was a predominantly rural area on the edge of Chelmsford in Essex, threatened by a major expansion of the town. The local authority forged ahead with this development in the 1980s and 1990s, but during construction work archaeologists uncovered what were considered significant finds including an elongated burial mound or 'curcus'. The mound and other features were entirely destroyed by the development as there were no mechanisms at the time for protecting the finds or revising the development proposals to reduce impact on the underlying archaeology. Arguably (or very clearly), planning was at fault, guiding new development but failing to value what the site already had. This returns us to the point raised at the beginning of this chapter. In retrospect, we believe that the planning system should have afforded greater protection to the historic landscape. But by the same token, how many features continue to be undervalued and destroyed today? And how harshly will future generations judge us for these apparently careless actions? The big difference today is that we have 'invented' the idea of sustainability: a light guiding us to judge contemporary actions from a future perspective, and a framework encouraging today's policy makers and implementers to think not only about money and profit, but also about community and environmental impact. The extent to which this happens in practice – and externalities are fully recognized – is of course a critical question. No-one believes that the planning system we have today is perfect, and the tendency to relegate environmental and communitarian objectives behind profit has not disappeared since the publication of the Brundtland report in 1987. The world remains much the same, though pressure to change it has undoubtedly grown. Continued neglect of the historic fringe is perhaps evidence of business as usual in land development, and avoidance of such neglect in the future is one objective of a multifunctional strategy of the type introduced in the opening chapter.

Such a strategy would undoubtedly draw on the planning system's more recent experience of dealing with archaeology and history. When the government moved to introduce Planning Policy Guidance Notes (the PPGs) in the late 1980s – because, it was felt, there was too little clarity and certainty in the way in which planning law should be implemented in practice – it took just two years to bring forward guidance dealing specifically with 'Archaeology and Planning'. PPG 16 was published by the Department for the Environment in November 1990. This Guidance aimed to set out the government's 'policy on archaeological remains on land, and how they should be preserved or recorded both in an urban setting and in the countryside'. It also gave advice 'on the handling of archaeological remains and discoveries under the development plan and control systems, including the weight to be given to them in planning decisions and the use of planning conditions. The guidance pulls together and expands existing advice, within the existing legislative framework' (DoE, 1990: Para. 1). It represented practical advice for local authorities that arrived too late for Springfield. It also stated that 'archaeological remains are irreplaceable. They are evidence – for prehistoric periods, the only evidence – of the past development of our civilization' (ibid.: Part A: Para. 3). The Guidance seems unequivocal on the value of archaeology: it tells us who we are.

Throughout the 1990s, local authorities across the country responded to the publication of PPG 16 by undertaking assessments of local archeological features so that appropriate guidance and policies might be written into local plans. In some areas, counties became involved in creating frameworks for the protection and conservation of archaeological heritage. There were, of course, the more obvious archeological features that everyone agreed should be protected: hence, Dorset County Council has long concerned itself with the best ways to deal with visitor pressures at Stonehenge. In 1999, Essex County Council in association with Kent County Council, English Heritage and The Thames Partnership published 'An Archaeological Research Framework for the Greater Thames Estuary' (Williams and Brown, 1999). Much of this area is likely to be affected by the proposed development of the Kent Thames Gateway (the report predates the development of building plans for the area, which are soon to be set out in the Regional Spatial Strategy for South East England) and it is extremely rich in archaeological features; maritime archaeology involving, for instance, London's Roman port; inter-tidal archaeology with the remains of settlements, pottery production and oyster cultivation; seawalls and flood defences; settlements of many different ages; and countless features of industry, mining and transport. This framework proposes a research agenda with five priorities for research (into the detail of the area's heritage), a management strategy and the development of the educational potential of the archaeology of the area, which is also in line with the government's intention to conserve the historic environment partly in order to enhance a 'sense of identity'. Indeed, going back to PPG 16, in 1990 this Guidance Note stated clearly that archaeological remains are 'part of our sense of national identity and are valuable both for their own sake and for their role in education, leisure and tourism' (DoE, 1990: Part A; Para. 6).

If the fringe has such remains, then these will clearly play a part in shaping how fringe areas are planned and managed: this is already apparent from existing policy (Ball and Brown, 2000; English Heritage, 2004a, 2004b). We stated at the beginning of this chapter that all landscapes have historical associations, sometimes more subtly and sometimes more obviously expressed.

Because of the nature of prehistoric archaeological features, their rarity and their association with places that have disappeared from modern maps, they will not be more abundant in the fringe than elsewhere. In contrast, some more recent historical features have a special place at the edges of towns and cities, because they are the legacy of an industrial past, and of economic restructuring in the twentieth century. It is to this more recent industrial past that we now turn.

The industrial heritage and modern re-use

Like older archaeological remains, more recent industrial heritage bestows a sense of national identity, perhaps (on Britain) as the first industrial nation, or a nation that has rapidly shed its industrial past during the later twentieth century. Part of the focus of this section is on re-use, but our concern with reusing old industrial or military buildings needs to have a caveat. Re-use often provides the means to conserve, but this conservation should be seen as an objective in its own right: with commercial or recreational re-use viewed as a means to an end rather than an end in itself. In the case of industrial archaeology, historic remnants often connect directly to the urban development that has its end result in the towns and cities that exist today: such remnants might include water pumping stations or disused aqueducts, old gasometers or reservoirs with their ornate Victorian dams. These items in the fringe landscape were out there to serve growing cities. Conversely, towns and cities sometimes grew to serve the industries that have since ended. In the seventeenth and eighteenth centuries, cotton towns sprang up next to the fast-flowing rivers needed to drive water mills; many have since been abandoned, fallen into dereliction, been demolished or turned into themed pubs or restaurants. A hundred years later in the nineteenth and early twentieth centuries, new towns and cities expanded close to mineral workings in the English Midlands, South Wales and in the north of England. Indeed, the 'old' mineral extraction industries of coal and iron ore have perhaps left the greatest legacy for the rural-urban fringe (Figure 3.2).

The arrival of these primary industries prompted the rapid growth of many industrial towns, in response to the need to house the workers directly employed in extraction, in related processing industries (such as steel making), or in the wider local service economies. Extractive industries were central to Britain's industrial revolution, with heavy industrial production of goods for domestic and export use having an insatiable appetite for coal, iron and steel. In the age before footloose industry, it was essential that there was a ready supply of labour at the point of production. East of the Pennines, Leeds and Sheffield grew rapidly during this period, finding new prosperity in steel and coal. Mechanization and steam power gave new impetus to the cotton industry to the west of the Pennines. Manchester and its surrounding towns had already provided a focus for this industry in the previous century, enjoying the benefits – essential for the cotton industry – of a wetter climate. With industrialization the towns expanded, forming a major urban industrial conglomeration by the early years of the twentieth century (Figure 3.3). But each of the towns

56   A historic fringe

3.2   Redundant industries, north-east Derbyshire (Countryside Agency, Andy Tryner).

3.3   The industrial legacy (Countryside Agency, David Warren).

within this conglomeration retained much of its previous identity, and unlike London, it remains relatively clear where one town ends and another begins, not because of dividing expanses of green space, but because the components of Greater Manchester are today separated by the remnants of their industrial legacy, by now redundant mills, spoil tips, rail marshalling yards, and all the detritus of their heavy industrial pasts.

Indeed, as Britain's heavy industry has been consigned to the past, those towns that were central to elevating the country to the status of the world's first industrial nation have experienced rapid economic and physical decline, not only in their cores, but also at their peripheries. Many northern towns have fringes characterized by abandonment: by redundant mineral extraction and derelict industrial landscapes. Old pit buildings, cotton mills, railway infrastructure, disused canals and tramways, chimney stacks: all red-brick features of an industrial past that has given way to light industries, storage, distribution and out-of-town retail housed in prefabricated white warehouses, lit in orange-neon and surrounded by dual-carriageways. But occasionally, this utilitarian landscape advertises its heavy industrial past through the re-use of old brick buildings for more modern purposes. Perhaps the simplest cases of re-use occur where it is decided that a particular building is safe enough and sturdy enough to host a new use: walls are re-pointed, roofs and windows are replaced, modern wiring and plumbing are introduced (in the best cases), access is provided and the building is ready to play its new role. This only happens when it is cheaper to renew than to build from scratch, or where public money can be found to ready a building for re-use. The first scenario is rare: prefabricated warehouses can be thrown up cheaply and quickly; conversion, on the other hand, is expensive. The second scenario is becoming more commonplace. Old mill buildings in the north of England are sometimes boarded up by local authorities in order to halt the slow process of decay, resulting from illegal access and vandalism. Basic remedial work is also undertaken to prevent further weather damage. The hope is that this 'stabilization' work will encourage re-use, especially if prospective occupiers are eligible for some sort of grant support. The Groundwork Trust has a history of working with local communities and private business on land and building reclamation schemes since the early 1980s. The aim of such schemes is not merely to update economic use, introducing new industries into buildings intended for heavy manufacturing, but to integrate revived economic function into broader regeneration strategies. The conservation of a historic resource is another basic goal.

This is also the case at Duisburg-Nord in the Ruhr region in northern Germany. Here, the intention has been to integrate industrial archaeology with recreation, education and enhancement of the natural environment. The scheme has been held up as a shining example of what can be achieved on a large, hitherto derelict and contaminated, industrial site. Buildings and other industrial structures have been re-used for a range of purposes, while 'monuments' of the past have been retained. Those who applaud what has been achieved at Duisburg-Nord point to the new habitats that have been created, to the archaeology preserved and to the new opportunities for outdoor recreation that have resulted from the scheme. But there is some concern that industrial 'theme parks' will spring up around too many former industrial towns, with the same kitsch features: pit-wheels painted red or blue and strewn across the landscape in iconic depiction, blackened buildings running art-house slide shows of 'nineteenth-century life', and restaurants serving hotpot on tables illuminated with miners' lamps. And what would this standardization of the 'heritage package' mean for the sense of national identity that policy makers appear so keen to promote? Samuel (2000: 107) takes a slightly contrary view, arguing that 'far from simply domesticating or sanitising the past, it [the heritage park] often makes a great point of its strangeness, of the brute contrast between now and then'. But that is not to say that heritage parks will not deliver a standardized package.

Continued urban expansion in many parts of Europe means that the fringe and its industrial heritage face the constant threat of redevelopment and in this context, some sort of positive action – of the type seen at Duisburg-Nord in Germany, or indeed at Colliers Moss in England – seems preferable to inaction (Figure 3.4). And the management of both these sites accords broadly with what we consider to be 'good planning' at the fringe, which we elaborate on later in this book. However, good planning is not about formulating a standard package and applying it ad infinitum. There is inevitable – and well reasoned – reaction against such packages, and few people would wish heritage parks to become as ubiquitous as Starbucks Cafés. The problem is that planning may start off with the aim of protecting the unique, but end up creating the mundane. With regard to industrial heritage, there is now an emergent backlash against painted pit-wheels. We will of course say much more about planning in the final part of this book, but would note here that prescriptions for conservation should really be avoided: planning should be about working with place and with the special qualities of place, and not about prescribing one-size-fits-all solutions. Although these German and English examples square with the concept of multifunctionality, bringing together and promoting economic, historical, ecological and aesthetic 'functions', they might not offer perfect models for replication elsewhere. The principles might be sound, but they have to be transferred sensitively.

3.4 Derelict industries and formal leisure, Silloth, Cumbria (Countryside Agency, Barry Stacey).

Doing 'something' at the fringe is sometimes essential, especially where there is a risk that inaction will encourage redevelopment that might destroy those qualities (including historical qualities) that deserve to be protected or promoted. The industrial heritage – and indeed the twentieth-century military heritage – of the rural-urban fringe is often under threat from development pressure spilling over from nearby towns and cities. This is the case in southern England (comprising the South East, East of England and London regions) where the planned growth of housing, associated infrastructure and employment, is projected to be substantial over the next twenty years. In the South East alone, the Regional Assembly (the body responsible for overseeing the production of a Spatial Strategy within the region) is projecting that between 500,000 and 610,000 new homes will need to be built in the period up to 2026. Many will be concentrated in what the government has dubbed 'growth areas' listed in its Sustainable Communities Plan (ODPM, 2003a). In many instances, housing extensions will be bolted onto existing towns, and concentrated in adjoining land that is considered to be 'brown-field' or PDL (previously developed land). It just so happens that, through historical accident, there is a concentration of World War II airfields and other military sites (including old anti-aircraft gun emplacements and barracks) in southern England, concentrated in this part of the country as the first line of defence against attacks on London. Suddenly (actually this has been happening for a number of years, although the Communities Plan has heightened pressure), there is a scramble to find new sites for housing that fit with the government's definition of PDL, allow local authorities to achieve the official target of achieving 60 per cent of new development on previously developed sites, and do not upset the local electorate by consuming green fields. Given these objectives, airfields appear ripe for redevelopment and the natural location for new housing (Lober and Gallent, 2005). The same is if course true of old industrial sites, which many people consider to be eyesores and infinitely more suited to redevelopment than adjacent agricultural land. Looked at in this way, the argument might seem somewhat perverse: is it better to redevelop sites merely because they have been developed previously irrespective of the potential preservation value, and should intensively farmed green fields be 'preserved', because there is a de facto case for preservation that is never articulated? The argument is grounded in the classic NIMBY view: if there was nothing there previously there should be nothing there in the future; if there was something there in the past we can legitimately replace that something with something else. This perhaps returns us to the argument set out at the beginning of this chapter: the more subtle heritage of the fringe is often not highly regarded or valued. Very few military airfields have been deemed worthy of protection apart from a handful made famous through film or through the actions of Churchill's 'few' in 1940. For the rest, one of two processes normally take hold: either they fall prey to urban housing extensions (housing encroaches upon the site gradually, residents complain about the whir of light aircraft, flying ceases, and the volume builders arrive to replace hangars, runways, and control towers with red-brick boxes); or, in some cases, airfields in the rural-urban fringe have been put to other uses (Bell *et al.*, 2001); hard runways are used for motor-sports, for vehicle testing, car storage, or for Sunday markets; hangars are used for warehousing and storage or light industry; some general aviation may continue; and other buildings may be re-used as aviation heritage centres or museums.

The first fate befell Ipswich Airfield in the late 1990s. Little now remains of its former use: its history has gone, and its historical association has been lost. But the story at Wellesbourne Airfield in Warwickshire is somewhat different. Stepping back for a moment, we should say that military sites are common in the rural-urban fringe, not only in the south, but across the United Kingdom, and especially those dating back to World War II. Airfields, in particular, were built close to the cities they defended (e.g., around the peripheries of London, Coventry or Birmingham). Today, some of these airfields have evolved into regional and national airports, but many others have since become redundant. Willis and Holliss (1987) have shown that in England alone, more than 500 military airfields were constructed in the twentieth century and by 1985, 72 of these remained in military flying use while 67 had transferred to civil aviation. Almost 400 sites, many of which are located in the urban fringe, were no longer used for flying (suffering the fate of Ipswich in many cases) and work by Bell *et al.* (2001) has examined this historical legacy, looking at old fields in Suffolk, Warwickshire and Yorkshire. Many airfields have demonstrated multifunctional potential: usually for one of two reasons. First, old hangars provide ideal buildings for new industries, and closed runways provide great venues for motor sports, markets and car storage. Secondly, a desire to retain flying operations sometimes means that owners will seek to cross-subsidize flying by leasing space for other activities. Airfields illustrate the kinds of relationships that frequently develop between the legacy of the past and the needs of the present. Returning to the case of Wellesbourne (a village of close to 10,000 residents) in Warwickshire, the airfield closed in the 1960s and has since gradually developed into a space for new housing and jobs, while also creating opportunities for other economic and sociocultural activities. Arguably, this pattern of re-use is the product of individuals grasping particular opportunities rather than any holistic vision, but it serves to demonstrate the nature of the historic fringe and how this may be linked – either by fortune or by design – to other fringe functions.

This experience of re-use is very different from the experience of former industrial areas discussed earlier. In the industrial cases, a conscious decision was taken to use public money to clean up the sites, install visitor centres and promote some form of limited economic rejuvenation. But in the case of England's airfields, the patterns of re-use – which are repeated across the country – are more usually products of economic necessity. There are exceptions: at Duxford, the Imperial War Museum was funded to establish a centre designed to educate and enthuse, celebrating Britain's military aviation history. But for the most part, old airfields now in private hands have simply responded to the changing economic and political climate. Large sites – in some cases 4 or 5 km² or even larger – cannot sustain themselves on flying alone: hence they diversify, with owners leasing hangars for storage and distribution, closing some runways and renting space for weekend markets, selling off portions of land for office or retail parks, leasing hard runways for car storage. All these activities can be found at Wellesbourne and countless other airfields up and down the country (Figure 3.5). This is not manufactured multifunctionality: it is organic. It is the product of economic necessity, but at the same time, results in a retention of historical features as well as the promotion of wider access and a new economic function. Airfields become more functional through this pattern of re-use and tend to avoid the kitsch, a few celebrate their heritage with propellers protruding from concrete plinths but there is no 'standard model' for airfield re-use. The same is true of old army barracks, which are also commonly found in the fringe.

In some instances, former army barracks and other buildings may be re-used for industrial purposes or for warehousing and storage. In this way, some interesting buildings – perhaps rich in war-time heritage – are preserved, though they may not be open to the public. Occasionally, barrack buildings and messhalls have been re-used as hotels or community centres. Both types of re-use are relatively common. On the edge of Salisbury, the company Salisbury Investment Castings moved into a former munitions storage facility several years ago. The facility had lain empty for some time since closure. In terms of size and access, the site was ideal and located on the edge of the city. However, there was much work to be done to the actual buildings and immediate surroundings before Investment Castings and other companies could move in. First, the top soil was removed across the entire site to a depth of 1 metre, to ensure that any stray munitions (bullets, mortar rounds and so forth) would not pose any future risk to life or limb. Second, the buildings themselves

3.5 Small general aviation airfield, Dunkeswell, Devon (Countryside Agency, Pauline Rook).

needed to be adapted for their new uses. Salisbury Investment Castings is in the business of precision casting for industry (metal parts for trains), the military (ammunition shells), and even the production of metal ornaments and jewellery. It utilizes heavy casting machinery that requires ventilation. The building that the company moved into was designed to withstand indirect bomb blasts and has a reinforced concrete roof roughly 12 inches thick. It was necessary to bore holes for pipes through this roof, and also to rearrange internal walls of similar robustness in order to accommodate the large and bulky machinery. Conservation through re-use is often not the easiest or the least expensive option. It can require considerable investment of time and money, and it is usually cheaper to erect new purpose-built structures. We said earlier that re-use is not the end-goal; rather the objective is to conserve for 'conservation's sake' or to preserve some sense of local or historical identity that is reflected in those remaining buildings and sites dotted across the country.

This is perhaps an ideal that business locating in historic buildings might not always share: their objective is to remain profitable. For some businesses, relocating to a red-brick Victorian institution on the edge of town might be a plus. Clients may prefer the leafy surroundings of an old asylum to the utilitarianism of a white box in freshly landscaped gardens, and of course, such a location might provide a more pleasant work environment for employees. More profitable businesses might well be able to justify additional expenditure (and some grant support is likely to be available, especially if organizations such as the Groundwork Trust are involved). But it is sometimes difficult to reconcile economic and conservation objectives. While airfields might illustrate a unique pattern of re-use that tends to result in conservation, other historical sites often require public sector intervention and need to be part of a conscious effort to re-use. When this happens, local authorities and other agencies need to weigh the benefits of conservation against costs. Financial costs may be considerable in the short term, but what of the long-term price of not conserving these historical features, and of essentially severing the link with the past?

We will say much more about the role of planning later in this book, but it is perhaps useful here to reflect briefly on the role of planning in 'place making'. We have already mentioned ideas of 'non-place' (Augé, 1995): arguably places without association fall into this category. But how is place 'made'? In Chapter 2 we considered Hoskins' (1955) view that landscapes are 'made', not through human design, but through an interaction of different processes – economic, social, political and demographic – over time. Through human habitation, England has been transformed from a prehistoric wooded landscape to a landscape of intensive farming and red-brick boxes. The impact of farming on the landscape has been substantial especially since the eighteenth-century enclosures and the intensification of subsidy-based farming from 1947. The economic and industrial revolution left its own imprint on the landscape, as did two World Wars, one Cold War, and the creation of a post-war planning system that prioritized containment and, more recently, urban intensification. Through these processes and events, a landscape has been made which has left a physical imprint. If we remove this imprint – either overtly through demolition and clearance, or indirectly, by failing to value and conserve it – do we 'unmake' the landscape? The easy answer is yes, but in reality, there's a narrow dividing line between unmaking and remaking.

In Christian Rome, marble from the Colosseum was removed and taken to build houses and churches. During the medieval period, there was no conscious effort to safeguard the remnants of Rome's imperial past. In Athens, Turkish occupiers stripped the Acropolis of stone to feed limekilns (fortunately, the marbles adorning the Acropolis were taken to London for safe keeping). And in North Wales, the Normans destroyed the Celtic castle at Deganwy and used its stone to build a new castle to the south at Conwy. Were these places unmade through a lack of desire to conserve? Or is the process of remaking actually a process of making? In this sense, the making of place is not a process that can be judged here and now, rather it has a retrospective quality. By this token, the stone from Stonehenge could be removed tomorrow and incorporated into a community centre in some nearby town and we might expect that in a thousand years' time our actions might not be judged too harshly. But in all these examples – the real and the hypothetical – there is a strong retention of history, not only the history that created, but also the history that modified: the fall of Rome, the sacking of Athens, and the Norman Conquest. Landscapes are organic; they respond to events and to change. Although it is difficult to compare directly the fall of the Roman Empire with the arrival of Salisbury Investment Castings at a former munitions storage facility, the two examples illustrate the process of change that continually makes and remakes the landscape.

In this sense, conservation as a primary objective is flawed. Re-use means the retention of historical association, but this re-use will eventually create its own sense of history and, in time, bestow richer meaning on the evolving landscape. When someone eventually writes the history of the twenty-first-century English landscape, they will see a diversity of social and economic processes reflected in the physical landscape: they will see change, but with an underlying sense of continuity, and hence the identity that the government pointed to in 1990.

Up to this point, we have concentrated on built historical features that are important in all landscapes, but that are often found in the rural-urban fringe. Before moving on to consider very briefly the recreational potential of the historic fringe, it is important to note that not all the historical legacy comprises bricks and mortar. First, the landscape of the fringe is often a product of past farming practices: this means centuries of ploughing and the laying of hedgerow. Currently hedgerows located in the rural-urban fringe receive less protection than those in the wider countryside. This is because the Hedgerow Regulations – introduced in 1997 – require permission only for the destruction of hedgerows that are growing in or adjacent to any common land, protected land, or land used for agriculture, forestry or 'horsiculture'. Hedgerows growing in or adjacent to golf courses, sports grounds and allotments – which occur frequently in the fringe – are excluded from this protection. This perhaps returns us – full circle – to the argument that a lower value is placed on heritage at the fringe, and more generally to the view that the fringe is a second-class landscape. Second, another area of re-use – which has been relatively successful from an economic perspective, but is problematic in terms of restoration and conservation – is the re-use of old quarries for waste handling and land filling. Landfill has been particularly important at the fringe because of the historical legacy of ex-mineral workings, discussed earlier in this chapter. As we have already noted, many of England's towns and cities grew up next to mines and quarries, drawing on proximity to raw materials to fuel their growth in the nineteenth and early twentieth centuries. But the types of minerals extracted today are quite different from those that dominated one hundred years ago. The rise of fuel minerals in offshore and remote locations means that the rural-urban fringe has lost much of its importance as a zone for mineral extraction; though this is not universally the case. The legacy of mineral working abandonment in fringes has been exploited in many waste strategies, with former quarries often viewed as the ideal locations for the disposal of domestic and other waste. Government policy (articulated in PPG10) suggests that waste management facilities be located in:

1. Industrial areas – especially those containing other heavy or specialized industrial uses, degraded, contaminated or derelict land;
2. Working and worked out quarries – landfill is commonly used in quarry restoration but there may be opportunities for other types of waste management facilities at some quarried sites;
3. Existing landfill sites – where composting facilities may be conveniently located;
4. Existing or redundant sites or buildings – which could be used, or adapted, to house incineration or materials recycling facilities, or composting operations;
5. Sites previously occupied by other types of waste management facilities; and
6. Other suitable sites located close to railways or water transport wharves, or major junctions in the road network.

All the above types of sites are found in fringes, reinforcing a view that the edges of towns and cities are ideal for waste management; quarries are marked out as especially suitable and because the fringe has a history of performing a waste management function, it seems likely that this will continue and expand in line with government policy.

In terms of historical association, rather different issues are raised here. It is clear that hedgerows add interest to the landscape of the fringe: they break up the monotony of intensive agriculture; they mitigate the visual impact of roads; and they can provide a route for rambling. But quarries turned to landfill? There are two competing views: quarries are scars on the landscape and land-fill can be the first step towards healing those scars. Or conversely, they play the same role as hedgerows, adding interest and – through the addition of water – provide a valuable leisure resource. Whatever line of argument is accepted, quarries are a historical legacy and evidence of those place-making processes discussed earlier (Figure 3.6).

But leaving this argument to one side, what should be done with them, if in fact something needs to be done? The Groundwork Trust's view is that using quarries as waste facilities can be seen as part of a land restoration strategy, and in the best cases, may lead to landscape improvements. It is not clear whether this occurs in practice, or whether quality gains mean that ex-minerals sites subsequently used for waste disposal eventually find a further use – perhaps a recreation or conservation function. But this is often the aspiration of Groundwork. The Trust tends to accept the premise that former minerals workings are scars on the landscape, made worse by landfill, but which can be improved through the addition of top soil and a programme of re-vegetation. But in reality, how practical is it to use ex-minerals sites for waste management, and for recreation and conservation? Much will depend, of course, on waste management practices and filtering processes (separating hazardous from non-hazardous wastes) and the amount of funding available for restoration. Ex-industrial sites – where landfill has not occurred – may prove more attractive to those involved in restoration efforts, especially as it may be easier to convince people that uncontaminated sites have greater subsequent recreational potential. Indeed, the Landfill Regulations (2002) cite the following pollution problems with landfill: emissions of odour or dust; ground and surface water contamination; wind-blown materials; noise and traffic (prior to completion of landfill activities); the proliferation of birds, vermin and insects; the formation of aerosols; and the risk of fires.

These problems do not appear to suggest that landfill is the ideal precursor to the promotion of recreation, though over time and with considerable

3.6 *Quarrying at the urban edge, Blackwater, Surrey (Countryside Agency, Joe Low).*

investment, it may be possible to clean sites, beautify them, and provide attractive environments for walking and cycling. But really, this is about wiping clean the slate, burying the past, and accepting a more practical and economic use for this aspect of the fringe's historic past. Arguably, a celebration of quarrying in the fringe would involve retaining the physical feature itself and preserving old cranes and machinery. But is this really viable or desirable? Cities are growing and as they grow, they are struggling to deal with waste. The fringe has, historically, served the city. Its history is built on this relationship. Times have changed and cities no longer rely on the mineral deposits that were once found in the fringe; and this has caused the relationship between city and fringe to take a new direction. This is a useful point to end on: though the history of the fringe is important – and integral to what the fringe is – fringes also have to function on a range of other fronts. Saying that conservation is important for conservation's sake in all instances is to deny the need for the fringe to perform a wider range of functions. The rural-urban fringe needs to retain its diverse relationship with the city: fossilizing the fringe may cause it to lose its identity to a far greater extent than failing to conserve or re-use all buildings and the entire historic legacy at all times. Above all, the fringe must continue to function.

So far, we have argued that the historic features concentrated at the fringe can pose particular challenges for long-term conservation and re-use. Seeking to halt change is not necessarily the best way to conserve either the features themselves on the fringe's sense of place, or to promote continued functionality. There are of course practical considerations too, including the cost of conservation and re-use, which we have only touched upon in this chapter. It is also the case that re-use may be a cynical act, preserving façades, but ignoring the importance of internal layout or material use. These are more specific conservation issues that are not dealt with here. But the key point that we wish to bring out so far is that the fringe is a functioning landscape and what we see in terms of historic legacy are the evidence of this function. Hence, there are few castles or stately homes, but there is a richness in industry and military function. This is the legacy of a dynamic landscape: prioritize the wrong type of conservation here and we risk destroying this dynamism, which is potentially more important in place-making than these fragments of the past. This is not to say that future function should not be built on the past. Re-use as a conservation tool has a special place at the rural-urban fringe.

The historical fringe and outdoor recreation

A final digression before bringing this discussion to a close: the fringe has significant potential for recreation, not least because it is a resource close to people. We have already discussed at length the industrial landscape of the fringe and touched upon the issue of recreation potential. We return to the industrial fringe below, but first, would note that some fringes in England have a considerable amount of historical landscaping and parklands that were formerly used by an upper class élite for hunting (Epping Forest – not always at the fringe but now encroached upon by London's relentless pre-war expansion – is one example, previously cited, of such a landscape) or other forms of recreation. These country parklands – areas of grassland dotted with trees and frequently associated with lakes and historic buildings – often occur in the rural-urban fringe as part of landed estates. These were often situated on the edges of towns, to provide easy access and some measure of seclusion. A handful of examples include Luton Hoo, Windsor Park, Tatton Park on the edge of Stockport, Haddon Hall outside Bakewell, Euston Hall on the edge of Thetford, and Broadlands on the edge of Romsey. Parkland landscape is potentially extremely valuable for recreation of various types, including walking, picnicking, playing games, nature study, swimming and paddling. The preferred landscape of the British is, according to a study in 1965, 'a calm and peaceful deer park, with slow-moving streams and wide expanses of meadowland studded with fine trees' (Lowenthal and Prince, 1965: 192). It is not hard to see why parkland should be attractive to visitors: this is a landscape specifically designed to provide society's most privileged with the kind of environment that they would find most attractive. These landscapes are not a ubiquitous feature of the fringe, but often coarser, less 'ordered' landscapes still provide a range of recreational opportunities.

Much could be said on the broader recreational, educational and health potential of the fringe, but here, we wish to focus briefly on the association of historical features with recreation, turning to these related issues later in this book. The ex-industrial landscape offers numerous recreation opportunities: canals are predominantly concentrated in the West Midlands and South Yorkshire and in counties like Warwickshire, where few other water areas are accessible to the general public and little land is available for free wandering, hence the opportunities canals provide for country walking in the fringe (and elsewhere) are extremely important. That said, safety is sometimes an issue. Derelict sub-urban and fringe canals are not always an ideal choice for an early morning stroll, often because these are untouched by regeneration programmes (unlike sections of canal truncating the urban core) and are not known generally as locations for walking or cycling. This raises a question that we return to in a moment: is it important to 'sanitize' areas and 'formalize' recreation in order to make places at the fringe safer? Reservoirs also have a recreational potential; they frequently occur in the rural-urban fringe, and sometimes the public enjoys access to them for a range of recreational pursuits. Thus at Ecup, in the fringe to the north of Leeds, there is public access to the edge by virtue of the existence of public footpaths and a bridleway; Ecup is also a designated site of special scientific interest. However, at Hanningfield Reservoir, south of Chelmsford in Essex, on the other hand, public access was found to be restricted in 1998 to views from a public road and a small permissive picnic site on the 12-mile circumference (Shoard, 1999: 94). The Essex Water Company had blocked an attempt by Essex County Council to secure public path status to an existing track down to and along the water's edge.

Flooded gravel pits are also a frequent feature of the rural-urban fringe, since, particularly in south-east England, many towns have grown up in river valleys with gravel deposits, such as the Stour Valley in Kent and the Kennet, Loddon and Thames Valleys near Reading. In 1993, a study of recreation provision in Berkshire was conducted in the context of whether the countryside was becoming more or less exclusive. Its author considered that the Berkshire countryside was becoming more exclusive and, in the context of the use of gravel pit lakes, wrote: '. . . the re-use of water resources on agricultural land and following gravel extraction tends to be for income-generating activities such as permit fishing and organised water sports, with very little benefit for unrestricted public access, except in the case of Horseshoe Lake, Sandhurst. This is unfortunate as water resources offer a highly valued environment to informal recreationists' (Skellern, 1993 as quoted by Shoard, 1999: 95).

Perhaps one of the best examples of public provision of recreation facilities in the rural-urban fringe has taken place in the Lee Valley Regional Park, which covers 15 square miles of land centred on a 25-mile stretch of the River Lee and disused gravel pits bordering it between Stratford in East London and Ware in Hertfordshire. A range of new recreation facilities both for sport and for informal outdoor recreation has been provided amidst old gas works, sewage works, power stations, factories and glasshouses. But the Lee Valley Regional Park sits at the opposite end of a spectrum to the West Midlands canals mentioned above. Although the Lee Valley hosts a range of informal activities, its status as a Regional Park suggests some degree of formalization of recreation potential. Pressure to formalize this potential exists throughout the rural-urban fringe. The Countryside Agency, for example, tends to see recreation potential as something that needs to be harnessed and made safe: thus the Agency favours formal programmes of land reclamation, the creation of Greenways though the fringe, improvement to existing facilities, education centres with good public transport access, and landscaped parkland with nature trails, all created through a process of master-planning which returned to vogue with the publication of the Urban Task Force's 'Towards an Urban Renaissance' in 1999. We are not seeking to criticize this vision – at this point – but would simply warn that there is a considerable risk of turning dynamic and interesting landscapes into kitsch theme (e.g. heritage) parks, pit-wheels and all. The Birmingham canals are an interesting recreational and ecological resource, quite wild and rundown in places. Accepting that they have to be made safe, how far do we go? Is there a case for a 'lighter touch', repairing broken lock-gates, making adjacent buildings safe, subtle signposting and access gates? Or do we go with the urban design vision of newly cobbled towpaths, the latest lighting scheme from Gothenburg, minimalist street-furniture, wine bars, loft apartments and sustainable fringe living? A key question is whether or not the potential of the rural-urban fringe can be realized while still retaining a sense of the fringe as somewhere special? Or is this important? When we visit the fringe, do we really want to be somewhere else, imagining that we have arrived at an urban park with all the facilities we might expect in a modern city? Do 'visions' – including that set out by the Countryside Agency and described briefly in Chapter 1 – value the intrinsic qualities of the fringe, or do they seek to transform it into something else? It is difficult to judge the Countryside Agency's vision: it clearly does seek a transformation of the rural-urban fringe, but also talks about the dynamic nature of the fringe, which should not be discarded.

The fringe has not always been a place for people, though there is clear leisure and recreational potential in some instances. Historically, it has been a functional landscape, serving or being served by growing urban centres. The fringe has a strong sense of history, though this is sometimes underrated. In recent years, policy makers have suddenly turned their attention to the fringe, claiming that this is a forgotten landscape, that planning needs to do something with it: to realize its potential. In this chapter, we have shown that part of this potential arises from historical association, which can be kept alive through a continual process of re-use, creating a sense of underlying historical continuity. But there is also a second argument that emerges from this chapter: that neglect of the fringe – by planners and policy makers – has made the landscape that we see today. It is an unplanned landscape possessing characteristics and dynamism placed at risk by the sanitizing effect of land-use planning.

Conclusions

'It is in the past that England's importance and glory are seen to lie' and this nostalgic way of looking at places has resulted in a 'habit of seeing landscapes through past associations' (Lowenthal and Prince, 1965: 204–205). This antiquarian thesis has as much currency in the fringe as elsewhere, a landscape that played a key role in the industrial revolution of the nineteenth century and in the defence of Britain a century later. But the same authors also made the point – four decades ago – that 'land must be used, not left alone' to comply with English landscape tastes, which tend to applaud neatness and to deplore ragged dereliction (ibid.: 200). In this chapter, we have argued that landscapes go through a constant process of being made and remade, connecting them to their past. Land is used, and not left alone: but not in the sense intended by Lowenthal and Prince. Theirs was a view that function – at least for the English – is less important than 'neatness' and that action is only positive if it regulates and regiments the landscape. This view is strangely similar to some current thinking on the rural-urban fringe, and stands in opposition to those – including Marion Shoard (2000, 2002) – who would like to see the processes that have produced the fringe left to continue. Hence, dereliction and decay can come to be viewed as place-making processes in their own right.

History – like landscape – follows a path of production and reproduction. This suggests that the re-use of buildings and land is part of a natural cycle that need not be resisted (though we might stop short of dismantling Stonehenge). In the fringe, this process has resulted – in some places – in a rich historical legacy that remains dynamic and ever-changing, and frequently adapts to the needs of the modern world. The industrial legacy of the fringe also provides recreational opportunities, though these might need to be managed sensitively. Arguably, there are too many heritage parks springing up across the UK and Europe, and through the use of standard models, these may contribute little to a sense of place at the rural-urban fringe. Likewise, the planning of hitherto unplanned (or less planned) areas is not always the best way to promote economic vibrancy, or other areas of functionality, in landscapes characterized by informality and lighter intervention.

4

An aesthetic fringe

Introduction

The aesthetics of the rural-urban fringe is a theme that we have already touched upon at a number of points in this book. The 'look' of a landscape is in part determined by the arrangement of natural and made-made features: the contents of the fringe were introduced in Chapter 1, and re-examined in Chapter 2, where the quality of recent development (an expression of the quality of planning) was brought into focus. We have also introduced the idea that landscapes are 'made' in a physical and in a social sense. In this chapter, we intend to delve more deeply into the aesthetics of landscape, dealing not only with physical expression, but also with cultural images and representations of the rural-urban fringe, in media and film and in the collective public psyche. Essentially, this chapter is concerned with 'aesthetics', a word frequently associated with the fine arts, but more broadly defined as 'the area of philosophy that concerns our appreciation of things as they affect our senses, and especially as they affect them in a pleasing way' (Carlson, 2000: xvii). The idea of 'environmental aesthetics' is closely allied to the notion of 'landscape', a Dutch creation that first appeared in the sixteenth century in reference to the pictorial representation of rural scenery. Later, 'landscape' also came to signify a more general notion of the countryside as a 'visual phenomenon' (Muir, 1999: xv–xvi). In more recent times, the word has taken on a broader range of meanings: in the field of geography – which is perhaps the most relevant here – particular 'landscapes' were initially associated with particular regions (mountainous, forested, and so forth), but today the word has been realigned with aesthetics and with more subjective responses to particular places (Bourassa, 1991: 2–3).

The central concern of this chapter is the aesthetics of the fringe in this broader sense, both its physicality and the way the landscape finds cultural representation. A landscape definition put forward by Daniels and Cosgrove (1988) stresses this dual meaning of the word, and suggests that landscape is concerned as much with 'cultural representation' as with 'physical ingredients':

72 An aesthetic fringe

> A landscape is a cultural image, a pictorial way of representing, structuring or symbolising surroundings. This is not to say that landscapes are immaterial. They may be represented in a variety of materials and on many surfaces – in paint on canvas, in writing on paper, in earth, stone, water and vegetation on the ground. A landscape park is more palpable but no more real, nor less imaginary, than a landscape painting or poem.
>
> (Daniels and Cosgrove, 1988: 1)

Although we begin by presenting a broad overview of land uses and the 'physicality' of the rural-urban fringe landscape, the principal focus of this chapter is on representation and perception. The fringe is frequently portrayed as an ugly, scruffy or anonymous landscape, and such representations shape collective perceptions of the fringe. One further objective of this chapter is to challenge some of the more stereotypical preconceptions, and to try to explain why the fringe has not always received fair treatment. Because the fringe is regarded as a functional landscape, regularly experienced at speed (see Chapter 2), a tendency to ignore the fringe has taken root, which has resulted in a growing ignorance of this landscape. It is also the case that the fringe can be described as 'unplanned', or at least less planned, than more formal urban or rural landscapes (a theme again touched upon in Chapter 2). A lack of consequent order has been equated by some with a lack of aesthetic appeal: more planning could beautify the fringe, ridding it its scruffiness, and hence, of its stigma. Towards the end of this chapter, we look at the question of 'beautifying' the fringe through conventional planning, which is essentially designed to bring the fringe – or any other landscape for that matter – closer in line with current aesthetic norms and landscape tastes. The chapter ends by contrasting two opposing strategies: remaking the landscape, through planning, in order to reflect these norms; or challenging 'normalization' and celebrating instead the difference that is commonly found at the rural-urban fringe.

The physical landscape

The rural-urban fringe is a unique landscape comprising a particular mix of land uses and activities, 'made' by a collection of processes that have impacted on the fringe in specific ways. Previous chapters have suggested that this is a man-made, functional landscape that was shaped by the industrial revolution, by the rapid expansion of transport infrastructure and related development in the post-war era, and by economic redundancy and consequent consumption. Although we have noted the existence of particular land uses and activities in previous chapters (particularly Chapters 1 and 2), it is necessary to cover some of this ground again, outlining the physicality of the fringe as a context for examining representation. Few of the land uses listed below occur exclusively in the rural-urban fringe, but areas of fringe do tend to contain a higher concentration of these elements than more traditionally urbanized landscapes or the countryside. These uses come under a number of categories, and in no particular order include:

- Service functions, such as sewage works, rubbish tips, car-breaking yards, gas-holders, railway marshalling yards, motorway interchanges; we have argued previously that transport infrastructure developed to serve a decentralization of population in the post-war period (Hall, 1989), while the arrangement of service functions is a result of the splintered politics of the fringe and political expediency of locating un-neighbourly uses away from residential areas;
- Commercial recreation facilities, such as go-kart tracks, quad-bike tracks, golf courses, private fishing and water-sports lakes, nurseries, garden centres, football stadiums and golf courses; all the trappings of a landscape of consumption, which has developed following a period of economic redundancy in the fringe, occurring in the early and middle parts of the twentieth century;
- Noisy or unsociable but non-recreational uses that depend on the presence of a large urban population, such as catteries and kennels; again the product of a consumer society and also the political acceptability of developing at the fringe;
- Allotments, travellers' encampments and caravan sites; an interstitial, transient landscape is perhaps the perfect location for transient uses, e.g. allotments tended to be bolted onto inter-war suburbs as part of the government's wartime 'plough up' campaign;
- Mental hospitals, although many of these have closed in recent years as part of 'Care in the Community' with the sites redeveloped mainly for residential or commercial use; now consigned to the history of the fringe (see Chapter 4), but originally located in the fringe to create space between the insane and the ostensibly sane;
- Certain types of retail establishment, notably farm shops, nurseries, garden centres, superstores, retail complexes and shopping malls; the latter are the products of New Right planning in the 1980s and early 1990s, the former are more opportunistic uses, exploiting the open space of the fringe, proximity to green-fingered suburbanites, or the direct (farm) marketing opportunities afforded by near-urban farms;
- Factories, offices, business parks, warehousing; exploiters of the dense transport infrastructure which has expanded in the fringe since 1945 partly as a response to the demographic decentralization (to satellites) of some urban populations;
- A growing number of educational institutions and district hospitals, as the institutions involved have sold valuable city-centre sites and moved to cheaper land affording them more space in the urban fringe; the economics of land markets – which display stark contrasts between inner urban and fringe locations – has encouraged some public sector decentralization to the fringe. Universities may, for example, move book storage or sports facilities to the fringe;
- Farmland, often scruffy and under-managed, but by no means always so; land speculation at the urban edge of the fringe, and the anticipation of containment strategies weakening in the future (see Chapter 2), may encourage some landowners to allow the degradation of land in the hope

of securing future planning permission. The fringe is also the destination for legal and illegal urban waste, the latter through fly-tipping;
- Equestrian centres and land given over to horse grazing; the availability of cheap land, and proximity to middle-class customers, makes the fringe an ideal location for what has been termed 'horsiculture';
- Country parklands – the core areas of usually privately owned landed estates; an historical feature of some fringes (see Chapter 4), or now in the fringe because of urban encroachment over the past hundred or so years;
- Unkempt rough land, which may include land that cannot be developed because it is too steep or too contaminated, or land with a developed function that has ended (or may be awaiting development); in the 1950s, this would certainly include Nairn's redundant aerodromes (1955, 1957), though most of these have now either disappeared under urban extensions or have come to host a mix of new uses.

In contrast, the fringe is not characterized by a large amount of housing: it is a 'landscape without community' (see Chapter 2). Major residential areas are confined to the suburbs, though urban extensions may eventually penetrate outwards, or new neighbourhoods may be integrated into the fringe, perhaps as 'development fingers' juxtaposed with green wedges. Where housing does occur in the fringe, it often consists of fragmented ribbon development along a road, or by a roundabout, or comprises isolated dwellings built to house workers connected with agriculture, forestry or game-keeping that have survived the disappearance of such uses. Housing estates may abut the fringe, providing a sharp urbanized edge giving way to those uses listed earlier. But the fringe is not merely a random assemblage of these uses that defies decoding: rather, uses and activities come together in specific ways to give the fringe a special character.

We have already suggested that the fringe is transitional in a number of senses, not least because it is truncated by motorways that encourage 'visitors' to experience the fringe at speed: perhaps the term 'transit landscape' would be more apt. It is also 'transitional' in the sense that it provides something between town and country, if it is accepted that town and country are different in objective (physical form) and subjective (experiential) terms. But the suggestion that the fringe is transitional in the sense that what begins as 'urban' progressively becomes 'rural' (in terms of density, land-use mix, architectural styles and so on) is deficient in one particular regard: it fails to acknowledge the uniqueness, and the special qualities, of the rural-urban fringe. This 'transition model' is flawed as it ignores the particular ways in which the fringe has been made: if processes act on space in a particular way, then it follows that a particular type of space is created. The transition model posits that rural processes (a switch to intensive farming in the post-war period, and subsequently to the consumption of rural space) simply lose ground to urban process (a transition from heavy manufacturing to services uses) closer to the urban edge. In reality, the fringe displays some of these patterns, but processes have acted on the fringe in more subtle ways (see Chapter 2). Indeed, the transition model might suggest that, for example, offices, factories and shops (the standard urban land uses) would show a gradual transition from concentration in urban areas to a lower concentration in the fringe and a very low density in rural areas. However, farm diversification in recent years, impacting on the countryside, has seen conversion of a great many farm buildings to business and industrial uses, defeating the idea of a simple transition. Perhaps more importantly, the fringe contains its own particular uses (including essential service functions: for example, electricity sub-stations, gas-holders, water treatment centres, sewage works, car crushing yards, rubbish tips, recycling centres, motorway interchanges, airports and railway marshalling yards) that display greater concentration in the fringe than in any other landscape. As well as the dirty and noisy essential service functions, these can include low-density office, retail and industrial premises including warehousing and distribution. At the same time, political processes impacting on the fringe particularly during the last quarter century mean that fringe areas often contain just as much industrial, office and retail use as town centres and sometimes more. From either an urban or a rural perspective, the transitional view of the fringe is clearly deficient.

The rural-urban fringe has a different physical fabric. As well as the special concentration of key uses noted earlier, it also contains many non-agricultural uses that are essentially responses to urban demands for recreation. Included amongst these uses are golf courses, equestrian centres, motorcycle training grounds, sports fields, garden centres and country parks. There are also non-service (in the strictest sense) functions that have been excluded from either town or country. In the past, this category has included mental hospitals, now it includes travellers' encampments and might, in the future, include asylum and immigration facilities. Land under agriculture is also different in the fringe: it is frequently more fragmented than farmland found in the 'countryside'. Sometimes it is also more run-down than elsewhere (for the reasons noted earlier). In addition, the fringe characteristically contains other open land that is apparently unused: this may mean that it was associated with development such as a manufacturing plant that has fallen into disuse, or it may be an ex-military site; alternatively, it may be land allocated for development that has not taken place. As this land is often left to nature it can become rich in both the size and diversity of plant and animal populations, but because it is unmanaged, it may appear unsightly to the tidy-minded and may not strike any chord with more 'accepted' landscape norms (Figure 4.1).

This description of the physical fringe again tries to covey a sense of the landscape with which we are concerned. Ingredients have been catalogued that may not be present in all fringes, but many will be present. Community centres are present in some urban areas but not in all; the lack of a community centre does not make an urban area any less urban. And by the same token, the lack of a sewage treatment works does not render a fringe less of a fringe. A critical difference in the physical landscape is the arrangement of uses. Whereas urban uses are laid out on a template of streets and squares, fringe uses have a far looser arrangement. Whereas urban use may be integrated and connected, fringe uses may be fragmented and disconnected, partly resulting from the splintered politics of the rural-urban fringe (Chapter 2).

76 *An aesthetic fringe*

4.1 An untidy landscape, Aylesham, Kent (Countryside Agency, Anne-Katrin Purkiss).

Indeed, the road pattern serving new developments such as superstores and office parks often consists of many loops off a main artery, each serving individual premises unlike the shared front road access provided by a High Street.

This then is the general physicality of the rural-urban fringe, the production of which has already been examined at some length in Chapter 2. Our intention now is to focus on representations of the fringe, which make reference to this reality, but which also draw on a long tradition of accepting or rejecting features of the landscape on the basis of a benchmark establishing what constitutes a 'good' landscape.

Representations of the fringe

A great many of the land uses catalogued above might be labelled 'functional'. Lowenthal and Prince (1965), in their seminal work on English landscape tastes, have argued that the English tend to reject functional landscape and have, instead, nurtured a fondness for the 'bucolic', the 'picturesque', the 'deciduous', and the 'tidy' (ibid.: 189). But 'what is immediately apparent about interfacial landscape is that it embodies the antithesis of the characteristics likely to appeal to the appetites identified by Lowenthal and Prince' (Shoard, 2002: 121). This aesthetic rejection of functionality – and of more industrial landscapes – is not merely an English peculiarity: instead, it has been viewed as a universal phenomenon. Adorno (1984: 69), for example, has argued that 'the impression of ugliness [in technological and industrial landscapes] stems from the principle of violent destruction that is at work when human purposes are posited in opposition to nature's own purposes'. In the rural-urban fringe, where industrial uses are often seen against a backdrop of what some perceive as 'unspoiled nature' (i.e. the countryside beyond), 'the principle of violent destruction' is perhaps more accentuated than in the city where 'nature' is hardly visible and where the 'urban' has created a whole new paradigm of aesthetic beauty (Figure 4.2). This is apparent, for example, in Hoskins' *The Making of the English Landscape* (1955), in which he attacks modern land uses, buildings and vehicles, not because of their own (lack of) aesthetic quality, but because of their interference with 'the immemorial landscape of the English countryside', and with 'those long gentle lines of the dip-slope of the Cotswolds, those misty uplands of the sheep-grey oolite, how they have lent themselves to the villainous requirements of the new age!' (ibid.: 231–232).

4.2 Motorways and power stations, M18 looking toward Dray power station (Countryside Agency, Simon Warner).

78 An aesthetic fringe

While Hoskins lamented the impact of the modern on conservative rural England, Nairn (1955; 1957) launched a similarly vitriolic attack on what he called 'subtopia' – ('visually speaking, the universalization and idealization of our town fringes') in which he argued that 'by the end of the century, Great Britain will consist of isolated oases of preserved monuments in a desert of wire, concrete roads, cosy plots and bungalows. There will be no real distinction between town and country' (Nairn, 1955: 365). At the heart of his criticism was the view that the fringe resembled neither town nor country: 'the crime of subtopia is that it blurs the distinction between places. It does so by smoothing down the difference between types of environment: town and country' (Nairn, 1957: 355). Essentially, landscapes that accord with neither of these norms are undesirable: the rural-urban fringe was not evaluated as a landscape in its own right, but according to what 'makes' the right sort of town, or pleasant countryside. In half a century of subsequent landscape debate, Nairn's assessment of the quality of the fringe has stood its ground: the idea that the fringe is just something in between town and country, with no intrinsic characteristics of its own (remaining 'transitional'), is expressed today in the view that the fringe is anonymous, lacks identity and corresponds to Augé's notion of 'non-place':

> Airports, train stations, port terminals, and transport routes, have become social places of a mobile society. Transitional landscapes accompany such places. Interfaces in-between city and countryside, located on the periphery and mostly experienced at speed, they are often disorientating: travellers, temporary inhabitants of sequential landscapes, have no clear idea where they are or where the city starts. Transitional landscapes with their lack of identity correspond to what Marc Augé identifies as 'non-places'. They constitute undesirable journeys, made out of necessity and not of choice, as they do not meet any (scenic or other) expectations.
>
> (Kamvasinou, 2003: 182)

According to Augé, motorways and supermarkets – icons of the fringe – are prime examples of 'the real non-places of super modernity' because they 'have the peculiarity that they are defined partly by the words and texts they offer us' (Augé, 1995: 96). Hence 'main roads no longer pass through towns, but lists of their notable features – and, indeed, a whole commentary – appear on big signboards nearby. In a sense the traveler is absolved of the need to stop or even look' (ibid.: 97). And in the same vein, 'the customer wanders round [the big supermarket] in silence, reads labels, weighs fruit and vegetables on a machine that gives the price along with the weight; then hands his credit card to a young woman as silent as himself – anyway, not very chatty – who runs each article past the sensor of a decoding machine before checking the validity of the customer's credit card' (ibid.: 99–100).

The fringe is home to the non-places, clinically described by Augé. There was an explosion of out-of-town shopping in the 1980s, and larger regional centres, retail parks and factory outlets have become the icons of the modern

Representations of the fringe 79

4.3 More out-of-town shopping, Wolverton, Milton Keynes (Countryside Agency, Anne-Katrin Purkiss).

fringe (Figure 4.3). Indeed, the 1990s saw the opening of two of the most spectacular shopping centres to date: the Trafford Centre on the outskirts of Manchester and Bluewater near Dartford in north-west Kent. The former is a 1.5 million square feet development, boasting a dome larger than that of St Paul's Cathedral, an indoor sports complex, multiplex cinema and a hotel (Lowe, 2000: 261, 265). If railway stations competed with cathedrals in the industrial era of the nineteenth century (neo-gothic St Pancras in London being one of the most obvious examples), shopping centres now appear to do the same in an era of mass consumption. The cultural significance of the overblown proportions and architectural references are poignant: shopping centres are not simply functional structures, reflecting practical needs such as easy access car parking and the convenience of having everything under one roof: they are also highly symbolic buildings reflecting the ideals of our time (which seem to revolve around shopping, and of course the car). Benjamin (1999) regarded the shopping arcades and department stores of nineteenth-century Paris to be of major cultural significance: 'for the first time in history, with the establishment of department stores, consumers begin to consider themselves a mass' he wrote, 'hence, the circus-like and theatrical element of commerce is quite extraordinarily heightened' (ibid.: 43). Benjamin's circus-like theatricality is perhaps even more striking in the modern-day Trafford Centre. With its kitsch versions of different continents and colourful pastiches of grand streets, it resembles a theme park that replaces High Street shopping with a controlled and sanitized under-one-roof experience.

But there is also anonymity in the kitsch: describing the 'standardized world, the non-place world of the 1990s shopping centre' on the edge of Stockholm, Peter Hall concludes that 'it is almost indistinguishable from its counterparts in California or Texas' (Hall, 1998: 878). Whether in social democratic Sweden, or in free-market USA, this commercial fringe landscape looks essentially the same. Filmmaker Jem Cohen has given an artistic interpretation of these anonymous spaces in a film called *Chain*, which can be seen as a representation of worldwide sprawl: 'the full power of Jem Cohen's feature film Chain doesn't hit until the closing credits, which reveal that the movie's anonymous landscape of chain stores and highway interchanges was shot in seven countries and 11 American states' (Winter, 2005); 'Chain takes as its subject and setting the homogenised inter-zones of privately owned public space – shopping malls, hotel complexes, theme parks – that multinational corporations have remade in their own global-branded image, letting regional colour fade to a concrete grey' (ibid.).

Other more mainstream films have used the derelict and run-down aspects of fringe landscape as a backdrop for illegal activities. Shoard (2002: 130) has suggested that 'the aura of excitement that goes with the apparent lawlessness of the "edgelands" has been exploited in such films of the 1990s as *Reservoir Dogs*, *Pulp Fiction*, *Se7en*, *Things to Do in Denver When You're Dead*, *Fargo*, and *The Straight Story*'. The predominantly negative representation of the fringe in these films and elsewhere should, according to Shoard, be challenged in 'artistic expression of the dynamism that the interface enshrines, rather than simply the decay and redundancy with which artists usually identify it' (ibid.: 144). However, in aesthetic terms it is sometimes hard to argue that 'dynamism' is more appealing as an artistic theme than 'decay and redundancy', which has a long history in artistic and literary expression. The aesthetic admiration for ruins of historical buildings, for example, goes back to romanticism and is still apparent in Bernd and Hilla Becher's (2002) photography of derelict industrial structures. The power of art to sway perceptions and to challenge established conceptions of beauty is undeniable. Contemporary hip-hop culture has, for example, replaced the predominantly depressing connotations of urban decay with more heroic associations of (black) underclass struggle or cool 'street credibility'. Negative representations of the fringe might also give way to something more positive and charitable. Shoard (2002) ends her essay on the 'edgelands' with a plea for a champion to appear, who can give the fringe the same recognition that William Wordsworth and Emily Bronte gave the moors and mountains, and John Betjeman gave the suburbs. Maybe that champion already exists in

Iain Sinclair, who, in *London Orbital: A Walk around the M25* (2002), has engineered something of a fringe epic. This literary journey around London's orbital motorway, the M25 – 'The dull silver-top that acts as a prophylactic between driver and landscape' (ibid.: 3) – gives a much broader picture of the economy, culture and aesthetics of the rural-urban fringe, than any of the films cited by Shoard, in which urban edges are depicted as ideal locations for illegal money exchanges or for hiding 'kidnap victims' (Shoard, 2002: 130). That said, negative representations of the fringe are not confined to Hollywood. In the UK, official descriptions of the rural-urban fringe in policy documents often perpetuate the view that the fringe is scruffy and accommodates nothing more than 'essential but un-neighbourly functions such as waste disposal and sewage treatment, and contains areas of derelict, vacant and under-used land as well as agricultural land and woodland suffering from a range of urban pressures' (DTLR, 2001a: Para. 3.24).

The fringe has a physicality, and a representation that feeds perception. This representation has been approximately sketched out above, though our referencing of landscape literature has not been exhaustive. The intention now is to examine potential responses to the issues highlighted: to seek physical change in the landscape, accepting aesthetic norms; and/or (the two are not mutually exclusive) to challenge these norms and in so doing, reshape aesthetic perception.

Changing landscapes

The landscape of business parks, commerce, retail and roads – ubiquitous features of the fringe – is often perceived as sterile, creating little 'sense of place'. Similarly, industrial remains fallen into dereliction may lend the fringe a scruffy, unkempt appearance. Architecture on the outskirts of towns and cities is generally regarded as anonymous, and of poor quality. Perceptions can of course be challenged, and we will turn to this 'option' in the next section. But before doing so, let us consider the possibility – that seems to be winning support today – of 'improving' the landscape of the rural-urban fringe, bringing it closer into line with more traditional landscape tastes.

Business parks, Ibis hotels, Travel Lodges, science parks and other developments commonly associated with the fringe employ a standardized type of architecture (again, see Hall, 1998), but as Adorno (1984: 69–70) has argued, beauty is often found in 'buildings that have been adapted to forms and lines in the surrounding landscape; or in old architecture where the raw materials for buildings were taken from the surrounding area, as is the case with many castles'. In this respect, a less standardized, more locally based architecture, could change the now predominantly negative perception of the fringe. Appleton (1996) has followed Adorno's line of argument, suggesting that architecture with local references, or utilizing local building materials, tends to be more popular than architecture that makes no reference to its location:

> Regional vernacular building styles in England, for instance, however much they may embody more widespread principles of design have

> through their use of locally available building materials, established recognizable styles, which have made such an impact on public taste that, in some parts of the country, planning restrictions forbid the construction of buildings in any substance other than quarried or reconstituted local stone. Thus masonry surfaces are suggestive of cliffs or rock surfaces such as might occur naturally in that particular environment.
>
> (Appleton, 1996: 152–153)

It might be considered rather simplistic to suggest that a change in building materials would result in a beautifying of the fringe: clearly, there is great technical and aesthetic complexity in ensuring that building's are 'in keeping' with their context. But some simple steps might contribute to a changed landscape, over the longer term. For instance, one way of making the fringe more 'appealing' might be to encourage a more locally or regionally based architecture (that could be extended to fringes). Bath, in Somerset, is an often-cited example of how a local building material, the Bath stone, has given the area's architecture (independent of style) a distinct local flavour. Even post-war tower blocks blend surprisingly well into the Roman and Georgian setting, because they are built in the same material as the historical town centre. New housing developments in the rural-urban fringe could also make clearer references to the local urban area in order to emphasize its links with the town or city.

Apart from being associated with poor quality architecture and anonymous commercial development, the rural-urban fringe is often perceived as a scruffy and un-neighbourly landscape. Although this perception is over-simplistic – the fringe is often diverse, multifunctional, and multifaceted (again, see Sinclair, 2002) – a large proportion of the land that surrounds our towns and cities is to some extent degraded or apparently neglected. However, closer inspection of policy, monitoring and evaluation documents shows that a wide range of landscape improvement schemes in the fringe are currently trying to address the 'poor quality' of the landscape. Many of these projects are focusing on the often degraded land around the motorways that lead into or bypass towns and cities. The 'A5 Corridor Initiative' in Staffordshire, for example, is:

> designed to improve the environment along the trunk road corridor. The aim is not to focus solely on impressions of the area when viewed from the A5, but to secure improvements which bring benefits for the many communities which lie along the road, particularly within the Staffordshire coalfield areas. The initiative combines a strategic vision with an implementation framework which allows for the definition and implementation of projects by a variety of different organisations working in partnership, often at a local level.
>
> (Land Use Consultants, 2001: 13)

Similarly, in the 'South East Northumberland Forest Park', the aim of a 'Greening for Growth' project has been to increase tree cover around the towns of Ashington, Newbiggin, Bedlington and Blyth, and within the A189 corridor, from around 7 per cent to 25 per cent. The intention is to transform a degraded rural-urban fringe into an attractive, accessible, well-wooded landscape. Another initiative in part designed to improve an access corridor through the fringe into an urban area can be found at Van Diemen's Land in Bedford, where the provision of a multi-purpose area of community woodland is supposed to form a more attractive and ecologically diverse setting for new development (Penn Associates, 2002). Community forests – which are examined at greater length in later chapters – have also aimed to improve the appearance of many fringe areas, providing more 'attractive' approaches to towns and cities. Research by Penn Associates (2003) has indicated that many of the fringes currently under the supervision of Community Forest Partnerships are considered to be of low environmental quality and are often associated with industrial dereliction that is thought to reduce aesthetic appeal and discourage the arrival of potential investors:

> The presence of derelict and degraded land (generally former industrial land but which can include neglected agricultural landscapes) is often regarded as a major factor in influencing people's negative perceptions of an area. Such land may be termed as being in 'environmental deficit', often as a consequence of the failure or cessation of former economic activity.
>
> (Penn Associates, 2003: 25)

The suggestion that negative perception poses a barrier to investment and economic rejuvenation has become perhaps the most persuasive argument in support of beautifying strategies. The report by Penn Associates is clear on this issue: the image of an area is significant in determining the ability of a region to attract and retain investment from outside; it may deter local companies from investing further money in expansion and development. For example, the 'quality of life' in a region is regarded as highly significant for a company's ability to attract and retain a work force, particularly in the knowledge-based economies, as well as in tourism. In the north west of England, approximately £25 million is being spent across the region through the 'New Approaches Project'. This Regional Development Agency-initiated project aims to improve the landscape, and hence the image of the region, along strategic corridors. The focus of these schemes tends to be placed on 'greening' visual eyesores and enhancing the main transport routes to regeneration areas (ibid.: 33).

What all these schemes have in common is that they try to 'improve' the quality of the landscape by cleaning up perceived eyesores and creating a more orderly and 'natural' environment. However, it could be argued that there is nothing objectively more beautiful about a community forest than a derelict industrial site. Projects of the type just noted present an ideological view of aesthetic beauty. The Countryside Agency – and its partner, the Groundwork Trust – have lent considerable support to the Community Forests programme, promoting a particular 'countrified ideal' of the rural-urban fringe, which is intrinsically hostile to urban land uses. Shoard (2002) has criticized the Community Forest programme (CFP) and argued that:

> Afforestation has emerged as the fashionable mechanism for transforming as much as possible of the interface into something more acceptable to polite society, and hiding as much as possible of the rest of it from view . . . this, it is imagined, is the kind of scene that will lure industrialists and prospective homeowners alike, to fill the grim world of the edgelands with happy workers and laughing families. All over Britain, disused quarries, old industrial land and other varieties of unkempt wasteland are to be turned into something more respectable and legitimate – woodland.
>
> (Shoard, 2002: 133–134).

Shoard's polemic (and fairly scornful) choice of phraseology – 'polite society' and 'laughing families' – reveals a strong ideological content in her critique that echoes Adorno's analysis of the bourgeois rejection of industrial landscapes in *Aesthetic Theory*:

> In naively condemning the ugliness of a landscape torn up by industry, the bourgeois mind zeroes in on the appearance of the domination of nature at the precise juncture where nature shows man a façade of irrepressibility. That bourgeois condemnation therefore is part of the ideology of domination. This kind of ugliness will vanish only when the relation between man and nature throws off its repressive character, which is a continuation rather than an antecedent of the repression of man. Chances for such a change lie in the pacification of technology, not in the idea of setting up enclaves in a world ravished by technology.
>
> (Adorno, 1984: 70)

Adorno presents a choice: to seek a 'pacification of technology' (bring it under control?) or to hide, and hide from, what we have created, 'to set up enclaves in a world ravished by technology'. In the schemes noted above, the choice is clear: technology and the modern is to be hidden behind decorative tree planting schemes. It would be naïve to suggest that in all instances, the apparatus of the modern world – roads, incinerators, power stations, flyovers and so forth – should be celebrated and held in high esteem. However, there are alternative ways of viewing these apparatus, which has been lost on those choosing only to value the conventionally pleasing aspects of the landscape. Is the world so limited, that only wood and grass find favour, and the only colour regarded well is green? The view that the landscape of the fringe is bereft of beauty or appeal should surely be challenged, and indeed it has.

Changing perceptions

Lowenthal and Prince's (1965) argument that the English favour a neat and tidy landscape is a persuasive one, but it is also over-simplistic: the same is perhaps true of the suggestion that the bourgeois naturally reject industrial landscapes. Bernd and Hilla Becher's (2002) photography of industrial structures – found on coffee tables across the world – bears testimony to an increasing fascination with the aesthetics of the industrial landscape, and it is worth remembering that many of its features – including canals – which were hitherto negatively perceived, are now often cherished. The change of perception is partly due to the change of use (from industrial transportation routes to places for recreation, in the case of canals) but shifts in perception are also a product of changing aesthetic representation. The post-industrial landscape – described by Hall (1998) and others – might be subject to a similar reappraisal if artists like the Bechers turned their attention to its hidden beauties and made it the subject of future work. Bowman (2004: 413) poses the following question: 'if Bernd and Hiller Becher were to document the typical forms of early twenty-first century post industrial landscapes, what would they photograph? It wouldn't be the mills, mines and cooling towers for which they are famous. Would it be malls, distribution centres and computers?'. Historically, it is also possible to find examples of how technology has been 'aestheticized' in a positive way and incorporated into utopian discourses around modernity (see Kaika and Swyngedouw, 2000).

The Eiffel Tower is a prime example of a functional, technological structure – which, according to the original plans, was to be removed after the Paris exhibition – that became a landmark widely perceived as uniquely beautiful. More significantly in the context of the fringe, the 'urban dowry' (the built environment that supports urban technological networks) may be perceived as aesthetically appealing, not because of any 'ornamental display' (as in the case of the Eiffel Tower), but because it carries the promise of 'a better future and more equal society' (Kaika and Swyngedouw, 2000: 129). Whereas the 'urban dowry' is largely hidden underground within the city, it is visible on the fringes in the form of pylons, transmission stations, water towers, purification plants, windmills and so on. These structures have – so long as they are seen as contributing to social, economic or environmental progress – a potential to be viewed positively, also from an aesthetic point of view. In the early days of motoring, before the realization that a surge in car ownership would lead to problems with congestion and pollution, many of the structures associated with road traffic were perceived positively. Indeed, Kaika and Swyngedouw (2000) reveal that petrol pumps were 'fetishized' in the 1920s by Louis Aragon in his 'Paris Peasant':

> Painted brightly with English or invented names, possessing just one long, supple arm, a luminous faceless head, a single foot and numbered wheel in the belly, the petrol pumps sometimes take on the appearance of the divinities of Egypt . . . O Texaco, Motor Oil, Esso,

> Shell, great inscriptions of human potentiality, soon shall we cross ourselves before your fountains and the youngest among us will perish for having contemplated their nymphs in naphtha.
>
> (Aragon, cited in Kaika and Swyngedouw, 2000: 129)

Aragon's comparison between petrol pumps and Egyptian divinities seems almost like parody in the era of global warming and the Kyoto Protocol, but as late as 1970, Fairbrother celebrated the sculptural qualities of road interchanges and the spectacle of traffic:

> Road interchanges are modern three-dimensional arabesques on a splendid scale, and on city fringes where they commonly occur are often the final triumphant slicing up of an area messily and irretrievably sub-rural. In industrial parks they could be superb decorative structures flowing with coloured cars like giant toys. Roads and traffic can be exhilarating like airports and railway stations, and the number of people who park from choice beside busy main roads proves that silence and solitude are by no means what everyone wants of their day in the open.
>
> (Fairbrother, 1970: 228)

Both quotations illustrate that notions of aesthetic beauty are not static, but subject to reappraisal and change: equally, they reflect broader social values. Thayer (2002) has created a framework of thought around the perception of landscape in which he identifies three social attitudes: 'topophilia', 'technophilia' and 'technophobia'. In this model 'topophilia' is represented by plants and animals, 'technophilia' by man-made things that are seen as something positive in the landscape (i.e. ornamental features, or those viewed as ecologically sound) and 'technophobia' by man-made things that have negative connotations. A nuclear power station (Figure 4.4) would – according to this model – be viewed as a feature of a 'technophobic' landscape whereas a wind farm (Figure 4.5) could contribute to a 'technophilic' landscape (though such a view might be subject to considerable debate). As Thayer points out: 'Telephoto views of modern wind turbines have been used in marketing as positive symbols of quiet power to sell automobiles (British Honda Motor Company) and Christian music compact discs (Petra)' (Thayer, 2002: 107). The reason why wind farms – in spite of their arguably negative aesthetic impact – are seen as contributing something positive to the landscape is simply that they are associated with clean energy production and environmental concern. Nuclear power stations, on the other hand, seem to epitomize everything that is wrong about the modern world. In the aftermath of the nuclear disasters in Harrisburg in the US in 1979, and in Chernobyl in the Soviet Union in 1986, nuclear power stations have become almost iconic symbols of the vulnerability of modern societies and the threat posed by radioactive waste. However, nuclear power stations have not always been seen as features of 'technophobic' landscapes: in the pre-Harrisburg era they were seen as symbols of technological progress, and the distinguished

Changing perceptions 87

4.4 Power station – negative connotations? Church Laneham, Nottinghamshire (Countryside Agency, John Tyler).

4.5 Wind turbines – positive connotations? (Countryside Agency, Mike Williams).

architectural historian Nicholas Pevsner was an admirer of their 'hyperbolic cooling towers' (Tandy, 1975: 125). The changing cultural connotations attached to a certain building type, in this case the nuclear power station and its cooling towers, illustrate how the perception of particular landscapes can change over time. Judging by current debates around cutting carbon dioxide emissions, nuclear energy production is increasingly seen as a relatively clean way of producing energy (particularly if the jump from fission to fusion can be successfully made), and it might be that the cultural perception of nuclear power stations is about to shift once again.

Debates surrounding landscape aesthetics, centring on the work of Lowenthal and Prince in the 1960s, have now moved on. Arguably, functionality can have both positive and negative connotations and it is too simplistic to argue that function and change add up to 'bad' landscapes. Thayer has illustrated that functionality has associated symbolism: green energy production is more likely to be viewed as a positive landscape attribute than facilities burning fossil fuel. Landscapes are the product not only of the physical 'items' – roads, power-plants, warehouses, waste facilities, houses, etc. – that create a particular vista, but of the meanings attached to these items. This is an important message for the rural-urban fringe: activities and land uses that are perceived negatively, if assembled at the fringe, will generate a largely negative image of the landscape itself, eliminating the possibility of the fringe being seen as a positive approach to a town or city, or as a desirable destination for recreation, or indeed as an area deserving special protection. The fringe need not have the traditional landscape features identified by Lowenthal and Prince (1965), but it will need to have features that society at the present time – and in the future – values and perceives as being generally positive. Arguably, it already has these features in the form of its 'urban dowry', and people can be educated to accept a different view of the functional landscape, appreciating shades of grey rather than only monotone green. But there will always be the knee-jerk reactions, and the obsessive behaviour that follows, sometimes manifest as pathological tree-planting. In *Aesthetics and the Environment* (2000), Carlson evaluates the 'eyesore argument': that the environment needs to be cleaned not because of any ecological impact littering might have, but because litter is an 'eyesore'. The notion of 'eyesore' is an aesthetic perception that can be linked to Susan Sontag's notion of 'camp sensibility'. In drawing this link, Carlson argues that:

> The dilemma is that we are divided between two conflicting ways of dealing with something that we initially do not aesthetically enjoy: one is to change the world such that the object of aesthetic displeasure is eliminated: the other is to educate people to change *their* aesthetic sensibilities such that the object, although itself unchanged, can be experienced as aesthetically pleasing. In short, camp's transformation of our experience of aesthetically displeasing objects yields an alternative to ridding the world of such objects.
>
> (Carlson, 2000: 139)

However, perception and response are also about image and role-play: how the respondent wishes to be perceived. In Sontag's view, camp is about theatricality. Similarly, Booth (1999: 69) has argued that camp is 'a matter of self-representation rather than sensibility'. and in that respect, a declaration such as 'aren't those pylons sublime!' is more representative of how the person who declares it wants to be perceived, than of his or her perception of the landscape itself (Figure 4.6). The argument, essentially, is that people are shackled to convention and that perceptions of landscape need to be challenged at a more fundamental level. Shifts in collective sensibility can appear facile if these are not backed up by a deeper collective understanding. Sieverts (2003) presents 'para-aesthetics' as one potential answer:

> Para-aesthetics indicates something like an aesthetics turned against itself, or pushed beyond or beside itself, a faulty, irregular, disordered, improper aesthetics – one not content to remain within the area defined by the aesthetic. The first syllable *para* in this context is read to mean by the side of, alongside of, past, beyond, to one side, amiss, faulty, irregular, disordered, improper, wrong.
>
> (ibid.: 92–93)

4.6 Pylons, East Riding, Yorkshire (Countryside Agency, Mike Williams).

Para-aesthetics, in the sense employed by Sieverts, is about appreciation of chaotic richness within the *Zwischenstadt* ('between cities'): it has a parallel in modern art which has moved on from standardized ideals of beauty, and is more ready to embrace chaos and ambiguity. The same modern art is also about viewing the world differently, about alternative perspectives and representations.

Conclusions

This chapter should not be read as a plea to 'save the fringe', nor are we suggesting that landscape interventions are unnecessary. However, interventions that seek to beautify the fringe are often guided by misrepresentation and by a desire to transform the landscape abutting towns and cities into a peri-urban theme park: Disney for the retail consumer and for the inward investor. Some of the improvement schemes will result in something 'objectively' and 'subjectively' better, but 'subjectively' only when judged against a narrow band of aesthetic correctness. We will consider the role of planning at the rural-urban fringe in far greater depth in Part Three of this book. For the moment, a key message here is that perceptions of the fringe have been shaped by convention and by negative representation. There is more than one way of viewing landscape: it is possible to value the functional (Thayer, 2002), and see beauty in disorder (Sieverts, 2003). Alternatively, it is also possible to dismiss the unconventional and the chaotic, and to seek 'normalization' through policy interventions. This is the choice before those seeking to manage the rural-urban fringe. It is not a simple choice with mutually exclusive paths: in reality it is a matter of striking the right balance. Go too far in one direction and the fringe may become a tightly managed theme park; but do nothing, and we risk bringing to an end those processes of change and transition that have made the fringe what it is today.

5
An economic fringe

Introduction

In Chapter 2 we suggested that many post-industrial rural-urban fringes have been subject to a transitional process stretching from production, through to redundancy and eventually to consumption. The era of production began with the Industrial Revolution which saw either industry growing adjacent to existing towns and cities, or towns springing up next to the primary extractive industries that fuelled Britain's economic expansion in the late eighteenth and nineteenth centuries. Subsequent economic shifts evident from the late nineteenth century onwards and growing international competition and economic differentiation (the essential ingredient of globalization) put many UK industries – coal, steel and various forms of heavy manufacturing – under intense pressure. Thus the pre-war and immediate post-war periods became characterized by a progressive abandonment of industrial sites located at the edges of many towns and cities. But other processes in the post-war period – the creation of green belts, the demographic outflow to satellite and new towns, and an expansion of peri-urban road networks – sowed the seeds of new economic opportunity in the fringe. These opportunities began to be realized from the 1960s onwards, first with the relocation of light industries to fringe sites (exploiting the locational advantages provided by new motorways), and eventually with a raft of uses including office space, storage, distribution and retail. And throughout this recent period of transformation and consumption, farming has maintained a position in the fringe. We have already sketched out the narrative of economic shift at the rural-urban fringe: in this chapter, our intention is to bring the economic story of the fringe into sharper focus, but also to concentrate on more recent (late twentieth century) changes. We revisit some of the earlier developments and processes already mentioned, and also introduce some new concerns that have not been touched upon so far – including the position of farming in the fringe and the continuation of some extractive industries. For the sake of completeness – and given our intention to build a full picture of the fringe – our focus in this chapter is the economic functioning of the fringe, and the future of the fringe's diverse economic base.

An economy in transition

In Chapter 1 – and in parts of Chapters 2 and 3 – we noted the fringe's role in serving towns and cities: in hosting the 'urban dowry'. This suggests that the fringe is part of the wider life (including the economic life) of urban centres; these centres could not function without the fringe, and more specifically, without those essential services that are provided within the fringe. But calling the fringe 'productive' suggests that it has an economic life of its own. The word 'productive' is perhaps misleading. It suggests that something is produced in the fringe and harks back to a time when such areas produced coal, heavy aggregates and steel to fuel industrial growth. It was suggested earlier that production has been relegated to the status of historical footnote, and that post-production abandonment has today given way to consumption. Clearly, primary production has been replaced by lighter industries and by emergent service sectors that require office and retail space: the fringe remains productive in the sense that it retains an economic function – which we illustrate in the remainder of this chapter – but the nature of this function has shifted in line with a wider economic transition. Two early points, therefore, can be made with regard to the 'economic fringe': first, this is a landscape that services a wider economy; and second, it is a landscape that possesses (and always has possessed) internal economic features of its own.

Its particular economic features spring from the role it plays as an interface between town and country. Historically, farming activity was at its most intense in the immediate countryside around towns. Without any means of rapid transport, and because it was in no-one's interests to see food spoil, farming took place close to urban markets, close to where the produce was needed. This necessitated intensive agricultural activity in the fringe, and therefore this use retained a dominant position. Hall et al. (1973) observed that at the beginning of the nineteenth century, urban areas remained 'neatly packed' within clear boundaries separating them from adjacent farmland. But industrialization – along with the development of better transport – changed the relationship between city and fringe. At first, extractive industries replaced agriculture in some instances, though a lack of mass public transport (in the first half of the nineteenth century) meant that cities remained compact, but with some blurring of industrial and residential functions at the inner edge. Second and more significantly, the development of cheap public transport – and the capacity to transport food over longer distances – resulted in a redistribution of the urban population and further spillage beyond historic urban boundaries (Hall et al., 1973: 76). The processes making the fringe – including transport development and economic transition – were detailed in Chapter 2: while a decentralization of people, first to suburbs and then to satellites, saw the fringe become a buffer between places (*Zwishenstadt* or 'between cities' to use the German label), economic transition was in part driven by the development of peri-urban arterial and orbital transport links.

Thus, the importance of transport in reshaping the economy of the fringe is critical: a lack of transport resulted in an intensely farmed fringe; a densification of transport infrastructure has made the fringe the ideal location for retail and services in the post-war period. And in between these extremes, a period of industrialization and urban growth made the fringe the ideal location for un-neighbourly but essential services. Of course, not all fringes are the same, and containment policies in the form of statutory green belts – examined at length in Chapter 9 – have prevented some fringes from welcoming emergent service industries or retail.

But for the most part, and despite a raft of planning restrictions that seek to promote containment and prevent development in open countryside, the new economy of the fringe comprises a mix of activity: mass retail centres, business and science parks, warehousing, light industrial uses, farming, recreation and essential service functions that serve broad regional and national growth. Drawing on the work of Guy (1998) and Lockwood (2001), we have already suggested that some of these commercial functions, which have gravitated to the fringe in recent decades, have been viewed as problematic. General dependence on road networks and cars for customers and workers is an obvious drawback; so too is the impact on town and city centres of decentralization of certain functions, especially retail. But on the other hand, some forms of bulk retailing (hardware superstores, garden centres, specialist wholesalers and so on) are often more suited to fringe locations, largely because such operations require considerable expanses of floor space that would be prohibitively expensive in many centres. Similarly, some specialist retailers (plumbers' and builders' merchants, large car showrooms, etc.) often favour fringe locations for the same reasons. Light industry shares this need for space and the growth potential afforded by a non-centre location. Loading and distribution are also more difficult in central locations, while the best fringe sites are preferred because they frequently have good access to the radial and orbital roads that have proliferated around towns and cities (serving population decentralization) since World War II, affording good access to customers. Warehousing and distribution are drawn to the rural-urban fringe for the same reasons: cost, space and access. And business and science parks have developed because they share similar concerns and priorities. Space and access are central to the new economy of the fringe. The fringe is not a dead landscape: the government has been at pains in recent years to point out that while towns and cities might be viewed as the engines of the national economy, what happens in the countryside and in rural hinterlands plays a part both in oiling these engines and in providing the space for economic diversity. In Chapters 1 and 3, we considered the example of reusing old airfields, many of which are located at the rural-urban fringe. In many instances, airfields have been transferred to a range of commercial uses: storage, distribution, retail (fixed and transitory in the form of open-air markets), light manufacturing and recreation. They are often dominated by a service economy, and present a microcosm of economic change at the fringe, moving from a single traditional activity (military or non-commercial flying) to a multitude of alternative uses through necessity and opportunity.

But the economic transition at the fringe is only partial in the sense that the emergent secondary and tertiary industries that have found a home in the fringe share that space with more traditional activities including farming and extractive industries and those other functions that service growth. It would be possible, in the remainder of this chapter, to look at each industry in turn, to describe its origins, its current state, and its future prospects. But our approach is to parcel the current economy of the fringe into three broader themes: (1) the new economy of the fringe, which is dealt with first and viewed in the context of interstitial transport development, and the fringes 'survivors'; (2) farming, which is yet to realize its full potential at the fringe, and (3) extractive industries, which are not always wealth-creating, but which become integrated with the fringe's broader service functions. The relationship between new industries and the survivors says much about the multifunctional nature of the fringe, and the prospect of building on and strengthening this relationship provides a further opportunity to comment on the future role of multifunctional planning and management strategies at the urban edge.

The new economy

Transport infrastructure and the 'profitability' of the rural-urban fringe

Transport development, motorization and the private car have all had a major influence on the development of the fringe, driving economic change, and subsequently resulting in sociocultural shifts (encouraging urban populations to access the fringe, occasionally for recreation, but nearly always for retail), and rapid ecological and aesthetic change: the landscape has been transformed, become degraded, fragmented or more interesting, depending on one's perspective. Given that transport occupies a central position in the fringe's economic history, it is perhaps useful to start with an overview of post-war developments. The pace of transport development in the nineteenth century was rapid in comparison with previous centuries, but it was nothing compared to the pace of twentieth – and particularly late twentieth – century change. Between 1911 and 1959, total road mileage in the United Kingdom increased by only 16,769 miles, or 9.5 per cent (Bagwell, 1974: 372). But between 1959 and 1970, motorway mileage increased tenfold while the number of road vehicles rose from a quarter of a million to over eight and a half million: a rise of 3400 per cent (ibid.: 367–368). Road building and road use are closely linked: roads not only absorb demand but also generate new vehicle use (presumably because new roads increase access to more places, and encourage those who previously used trains and other modes, to switch to cars). Because of the population movements described by Hall *et al.* (1973), the densification of road networks in the fringe was particularly pronounced. Roads also fuelled greater movement and came to be identified with post-war new towns. Urban growth and the need to access satellites or different sections of cities more rapidly, or to avoid congested urban road

The new economy 95

systems, also provided the impetus for the development of orbital roads and bypasses, with their links to arterial motorways and trunk roads. The early 1960s saw a huge expansion of the UK's road network, which initially fragmented the fringe, but which subsequently became a framework for economic decentralization and the emergence of a new fringe economy. Road building not only fuelled traffic growth; it also created accessible and profitable locations for this new economy. The north west of England and the fringes surrounding the Greater Manchester and the Mersey Belt provide illustrations of this process.

The north west of England has one of the densest motorway and trunk-road networks in Europe, centring on the southern portion of the region (see Figure 5.1). Regional Planning Guidance for the North West (Government Office for the North West, 2003), indicates that the region has a total of 14 long-distance strategic routes (5 classified as trans-European network routes), 17 strategic access routes, and 10 other routes of regional significance.

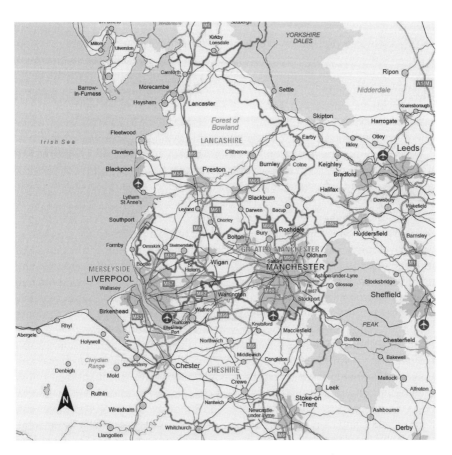

5.1 Major motorways and roads in the north west of England. (Source: Government Office for the North West [Graphics Unit 1/05]; Crown Copyright, based on the Ordnance Survey Map – Licence Number: 100018986.)

To the south of the region, major arteries (including the M6) pass between Liverpool and Manchester; these two major centres have orbital roads connecting their outlying parts (places like Bolton, Wigan, Bury and so forth in the case of Manchester), and an interstitial network of trunk roads and motorways connecting other secondary settlements (see the north west of Manchester in Figure 5.1) and the region's ports and airports. The north west has experienced major problems in recent years in terms of the 'hollowing out' of inner urban cores and a collapsing housing market in some areas (in central parts of Liverpool, western parts of Manchester including Salford and the area to the immediate east of Manchester's city centre). In this economic context (with the north of England lagging some way behind the south in terms of productivity, economic output and house prices), the motorway network has proven to be both a blessing and a curse. A blessing because it makes the region attractive to inward investment: it has well developed infrastructure including international airports, port facilities and a road network that affords easy access to locations within the region and access to London and the south. But a curse, because development has tended to concentrate at transport hubs, and especially where access routes to the south meet the regional motorways. This has created a number of difficulties: first, there is now a daily misery-go-round of commuting to and from these hubs for employment, shopping and leisure (Figure 5.2); second, some of the hubs have attracted housing development, helping to weaken inner-city markets in the poorest parts of Manchester and Liverpool; and third, much of the development has been of very low quality including the incredibly kitsch Trafford Centre (retail) and the proliferation of red-brick boxes masquerading as homes.

Motorway hub development in the fringes of Manchester, Liverpool and many of the other region's smaller towns has become a real headache for local authorities. They want to encourage inward investment and housing renewal, but wish to direct it to the hollowed-out cores. But the higher costs of redevelopment in these old-industrial centres and reduced access advantages mean that authorities are forced to either accept the preferences of investors and developers, or see development go elsewhere within the region. But further development of the fringe economy (of retail, distribution, storage and so forth) is clearly less problematic than housing. The new developments that have sprung up in recent years are well connected to roads, but completely disconnected (by distance) from services, jobs and nearby towns. In some instances, such developments present a potential body-blow to the government's Housing Market Renewal Initiative (HMRI: part of the 2003 Communities Plan (ODPM, 2003a), with authorities seeking to restructure housing markets in inner areas (by rationalizing stock), but then finding that these restructured and rationalized markets are having to compete with new edge developments. In this instance, the fringe is competing directly with nearby towns and cities for jobs and for people. This situation is fairly unusual. One of the problems in the north west has been that many local politicians in the region have been seduced by investment and have made decisions contrary to the advice of officers, permitting developments that go against market renewal and economic regeneration strategies. But, of course, some

5.2 Surface car parking associated with retail, Chieveley, Berkshire (Countryside Agency, Anne Seth).

of the developments are more suited to the fringe and enjoy economic advantages that cannot be secured in inner areas. There is a strong case for claiming that the Trafford Centre on the M60 should never have been built – according to Bryman (2004: 64), it is another example of the 'Disneyization' of mass consumption – but clearly it only works today because of the huge amount of car-parking space and easy access that its Dumplington location permits. Similarly, the clustering of light industrial and retail uses on the A34 has been attracted to the fringe because of easy access to Manchester International Airport and the abundance of cheaper land in this edge-of-south-Manchester location. Any future expansion of this network (or indeed similar networks elsewhere) is likely to fuel further edge development of the type that has been archetypal in the fringe between the north west's major urban centres.

But it is not only road development that has encouraged a proliferation of new economic uses in the rural-urban fringe. Rail links also provide a relocation incentive for some businesses, not only because of access to freight rail – though this can be a major factor for manufacturing industry – but because the recent emergence of 'parkway' stations has created new dynamic transport hubs. For retailers, an opportunity exists to exploit the commuter market, encouraging those returning from work to shop on their way home. For offices, a parkway location affords easier access to clients. The somewhat antiquated image of urban edge rail is of expansive and unattractive freight terminals incorporating maintenance depots and marshalling yards. Although these are still commonplace in the fringe, the new out-of-town rail hub frequently comprises a new white and glass parkway station situated in landscaped grounds,

98 An economic fringe

with ample parking and located within a larger complex of retail, factory outlet stores and office space. There are many examples of such developments dotted across the country. Commuters drive to the nearest parkway in the morning (avoiding the need to head into town to catch a train), park their car, and continue their journey to Glasgow, London or Birmingham by rail. In the evening, the journey is reversed. They arrive back at the parkway between 6 and 7 o'clock and then have an opportunity to shop at the Tesco hypermarket or spend an hour in some giant gym and squash complex, before completing the trip home. Heading into town for work, shopping, or leisure is now a thing of the past. Those still travelling by car south along the M1 into London are now bombarded with advertisements for the revamped Luton Parkway, which provides an entry point for fast services to King's Cross or Thameslink services into the city. The same opportunity exists for commuters to Birmingham who can now leave their cars at Warwick Parkway (privately constructed and opened in 2000) or for Cardiff-bound commuters who can use the now hugely expanded Bristol Parkway station (Figure 5.3: originally opened in May 1972). These are all rural-urban fringe stations, located many miles (in some instances) from urban cores, which all serve essentially the same function: servicing the needs of commuters.

Parkways of course are not separate from fringe road networks; rather, they are a further development of these networks: a response to increased road traffic, and an attempt to instigate a 'modal switch'. However, there is little evidence that parkways – or more general park-and-ride schemes – lead to any reduction in road traffic. In fact, it has been observed that park-and-ride and the expansion of parkway facilities can generate additional trips and

5.3 Bristol Parkway (Peter Lawson).

sustain car dependency. The Bristol Railway Archive describes the new Bristol Parkway as 'groovy', but the parkways have a large amount of surface parking. Parkhurst (2000: 160) has suggested that 'most people would perceive that the fringes of urban areas are more pleasant without the presence of surface car parks than they are with them. The critical policy question is not whether the provision of car parks is desirable, but whether it can be justified by economic necessity and/or traffic policy'. On the issue of additional trip generation, two points can be made: the first is that the development of parkway stations may encourage people who previously worked locally to take up jobs further afield. Average trip length at a regional or intra-regional level may be extended. Second, parkways generate new economic opportunities by virtue of increased accessibility. New retail and office space may cluster around parkways. If this is new inward investment, it may be deemed good for the local economy; if it is simply a relocation of existing local activity then it may be judged problematic, especially if journey lengths are stretched and car use increased.

It is probable that businesses clustered around the parkways – or other transport hubs – derive from both local and external investment. Transport development at the fringe has not merely wrested development away from towns: it has created opportunities for new types of retail; for businesses that need adaptable space; for activities that depend on rapid access to clients and customers across the country or the globe; and for businesses that have been priced out of cities, or that require low ground rents in their initial start-up phase. It is perhaps not too fantastic to describe some areas of fringe as the incubators of a new economy, offering the locational advantages and low costs that many new, perhaps experimental, businesses need. Issues of cost and accessibility make the fringe a potentially profitable location for many economic activities.

The relocation and emergence of retail space in the fringe was an issue touched on in Chapter 2. It is also an issue that has received considerable attention in planning literature (see, for example, Guy, 1998; Hall, 1998; Jackson, 1999; and Lowe, 2000). A concentration of retail space in the rural-urban fringe has had a neutralizing effect on edge landscapes (see Chapter 3), with Hall (1998: 878) arguing that shopping centres and other uses have resulted in a 'commercial urban fringe landscape' that is almost indistinguishable from one country to the next. But it is not only this aesthetic levelling that has proven controversial with the movement of retail to the fringe. As well as concerns over increased car use, there has been a long-running debate over the deadening effect that retail can have on more traditional shopping centres. The Trafford Centre mentioned earlier is not home to bulk retailers, but to a full range of High Street stores. It also tries to mimic a traditional High Street with a mock market selling specialist food, groceries and 'local produce'. In the 1970s, such retailing in off-centre sites was unheard of, with the focus remaining on 'bulky' goods and DIY; in the 1980s, there was an increase in electrical retail at the fringe; in the 1990s, the flood-gates really opened and all manner of retail became permissible within large retail centres (Guy, 1998: 306). But retail impact studies have reached different conclusions. A study of the Brent Cross shopping centre in North London found that 'by 1978 the effects of the Brent Cross shopping centre had been spread widely so that no individual centre had been severely affected' (Downey, 1980: 10). A study of the Metro Centre in Tyneside in the North East of England, on the other hand, concluded that 'the Metro Centre took a substantial share of retail sales, second only to the traditional regional centre. Its impact was not, however, focused directly on that city centre but spread widely across its broad catchment area' (Howard and Davies, 1993: 148). The out-of-town shopping debate has been running for at least thirty years, and can be traced back to the expansion of orbital and arterial roads in the fringe. Under pressure from local authorities, central government has grappled with this issue through a periodic review of planning policy guidance (PPG Note 6) relating to retail. Planning policy has been concerned with limiting the scope of retail on off-centre sites and with tightening restrictions on the development of new out-of-town shopping centres. The most recent revision of policy guidance – in the form of a new Planning Policy Statement 6: Planning for Town Centres, issued by the ODPM in 2005 – seeks to create 'vital and viable town centres' (ODPM, 2005: 5) by focusing development in existing centres. More particularly, PPS 6 states that 'all options in the centre (including, where necessary, the extension of the centre) should be thoroughly assessed before less central sites are considered for development for main town centre uses' (ibid.: 16: Para. 2.44). This appears to be a clear departure from the more 'liberal' approach to retail that developed through the 1990s, and the government now appears intent on supporting more traditional retail centres by requiring a 'sequential approach' in site search, where potential retail locations are considered in the following order:

- First, locations in appropriate existing centres where suitable sites or buildings for conversion are, or are likely to become, available within the development plan document period, taking account of an appropriate scale of development in relation to the role and function of the centre; and then
- Edge-of-centre locations, with preference given to sites that are or will be well-connected to the centre; and then
- Out-of-centre sites, with preference given to sites which are or will be well served by a choice of means of transport and which are close to the centre and have a high likelihood of forming links with the centre.

(ibid.: 16: Para 2.44)

Fringe locations are the least favoured in the government's new pecking-order. Parkway sites with access by rail, and usually with good bus connections, may well become a preferred fringe site for retail in the event of there being a lack of available space in nearby town centres. But generally, the sequential approach is recognition that fringe retail has been out-competing High-Street retail in recent years, offering a full range of goods to increasingly car-dependent customers. PPS 6 is government's belated effort to redress the imbalance and protect High Streets. But transport hubs within the rural-urban fringe are becoming increasingly accessible, with many having rail and bus

5.4 Yet more out-of-town shopping, Bradley Stoke, Bristol (Countryside Agency, Graham Parish).

links to nearby towns. The sequential test is likely to confirm the need for out-of-town sites for many retail activities, whose profitability depends on ensuring accessibility by car (Figure 5.4). Despite excellent bus and underground links to the Ikea stores at Croydon (via the south London tram-link), Edmonton, Thurrock and Brent Cross, few people choose to struggle home on the bus with the latest flat-packed wardrobe. Cars are an essential part of the sub-urban (edge-of-centre) or fringe (out-of-centre) retail experience.

Clearly, government is cracking down on some forms of fringe retail; however, this type of economic activity in the fringe is likely to remain vibrant in terms of mergers and acquisitions as aggressive operators seek to grow, and existing sites within the fringe change hands. Supermarkets have already responded to planning restrictions – and to market shifts – by opening additional smaller outlets either in High Streets or at transport hubs, sometimes in partnership with petrol companies. And while PPS 6 addresses competition between High Streets and out-of-town centres, the large supermarket chains have expanded their e-shopping and home delivery services, sharpening their competitive edge and maintaining wide profit margins. In April 2005, for example, the supermarket chain Tesco – which operates in 13 countries worldwide – announced annual profits exceeding £2 billion, up more than 20 per cent on 2004. Retail in the urban-fringe is already well established: supermarkets such as Tesco have diversified into traditional High Street merchandise including clothing and electrical goods. Government is acting to limit further expansion of these activities but retail – particularly retail of bulky goods including furniture and DIY – will remain a key economic activity at the fringe. Fringe sites are well connected; they tie in with commuting routes; they are profitable in terms of low rents; they provide ample space for expansion and for car-parking. The fringe has been able, in many instances, to respond to economic restructuring and the transition to a service economy, to a far greater extent than inner urban areas. Nowhere is this more apparent than in the north west of England where many old manufacturing cores have declined while new suburban and fringe hubs have rapidly become the focus of an emergent retail and office-based economy. The expansion of these new economic hubs may present a direct challenge to old cores, especially where they involve a diversity of functions and some degree of self-containment.

Edge cities of the new economy

In 'Beyond Suburbia', Fishman identifies a new type of urban form located beyond the suburbs, which he dubs the 'technoburbs':

> By 'technoburbs' I mean a peripheral zone, perhaps as large as a county, that has emerged as a viable socioeconomic unit. Spread out along its highway growth corridors are shopping malls, industrial parks, campus-like office complexes, hospitals, schools, and a full range of housing types. Its residents look to their immediate surroundings rather than to the city for their jobs and needs; and its industries find not only the employees they need but also the specialized services.
>
> (Fishman, 1996: 485)

This description of a typically North American landscape shares at least some similarities with the inner fringe that surrounds many towns and cities in the UK. The shopping malls, office complexes and industrial parks are all present – and evidence of the emergent fringe economy described earlier – though housing development in the UK fringe is often more limited, and few schools and hospitals have been decentralized to the fringe. Indeed, what Fishman calls a 'viable socioeconomic unit' in North America is more often a half-contained embryonic semi-urban unit in the UK. Johnston (2000: 14–15) has observed that the 'new generation of out-of-centre and out-of-town business parks are adding in leisure facilities, crèches, hotels, local shops and conference centres. Some, like the Kingshill scheme at West Malling, Kent, are metamorphosing into fully-fledged new communities incorporating large numbers of new homes'. Although this full socio-economic metamorphosis remains rare, Lowe (2000: 265) suggests that it may become more commonplace in the future. Bluewater in Kent (essentially the latest addition to Britain's collection of out-of-town shopping complexes), for example, is viewed by its Australian developers as a new town which is likely to expand its housing and business function in the future. Lowe describes it as a 'private sector city of the future' (ibid.: 265) and asks whether 'what we are witnessing now in the UK with the regional shopping centres is the development of edge cities like those charted by Garreau in North America?' (ibid.: 272). Garreau's (1991) 'edge cities' comprise new economic hubs that have expanded their range of functions, competing with and often out-competing traditional cores. In Garreau's rather prescriptive definition, edge cities have at least 5 million square feet of leasable office space; 600,000 square feet or more of leasable retail space; more jobs than bedrooms; they are perceived to be 'mixed use' destinations that 'have it all', from jobs, to shopping, to entertainment; and they were nothing like a 'city' as recently as 30 years ago. Lowe points out that Garreau's measures are difficult to replicate in the UK, though many new economic edge-centres are emerging in the fringe: these are seen by many as 'leisure centres' where superstores are complemented by a range of health clubs, restaurants and nightclubs. At the same time, 'business parks, distribution depots and housing estates also spring up in the interface, often around the bypasses and motorway interchanges that it provides' (Shoard, 2002: 123). A key question is whose responsibility it is to provide public amenities in these private-sector new towns. Is it the private developers or the local councils? In some areas, local councils have chosen not to provide services:

> there is, of course, a fear that provision of public facilities in the interface would further hasten the depletion of town centres which is already so pronounced. Yet the absence of basic facilities either encourages people to travel, usually by car, if they have the time, or it leaves them stranded in conditions we would not tolerate elsewhere.
> (Shoard, 2002: 131–13)

There is a long tradition amongst government at all levels, of resisting economic shifts in an attempt to defend places whose locations – ideally suited to Victorian or pre-war economic conditions – now render them less competitive. It is only natural, of course, that we should wish to protect historic cores, rich in heritage, rather than letting these cores stagnate and accepting the market processes that have resulted in some fringes becoming new centres of the service economy. There are many other reasons – social and environmental – for not simply letting the market determine spatial reorganization. But there is also a case for exploiting the locational advantages of the fringe for economic development. There is a reality that cannot easily be resisted. We noted in Chapter 2 that 'the coal mine and its neighbouring iron foundry, belching forth black smoke into the sky' described by Castells and Hall (1994: 476) has today been replaced by the 'campus-life atmosphere' of the contemporary business park 'set amidst impeccable landscaping' (ibid.: 476) and often located on the urban edge. Of course, some uses do not fit with this image and have been discouraged from locating in the fringe. The last Planning Policy Guidance Note 4 (DTLR, 2001b: currently subject to review) calls on planning authorities to ensure that most industrial development is directed to locations that minimize car use, and certainly away from trunk roads, where it might contribute to traffic congestion (ibid.: Para. 10). On the other hand, many uses are acceptable at the fringe if 'energy-efficient modes of transport' are available. Indeed, 'offices, light industrial development, and campus style developments such as science and business parks likely to have large numbers of employees' (ibid.: Para. 10), necessitate good transport

104 An economic fringe

links, perhaps of the type found at parkway stations (see earlier discussion) (Figure 5.5). Some uses are actively encouraged:

> Extensive, well-planned out-of-town distribution parks can offer economies of scale and consequent benefits to consumers or businesses supplied. Sites for such developments are best located away from urban areas, where the nature of the traffic is likely to cause congestion, and wherever possible should be capable of access by rail and water transport. Such sites should be reserved for those warehousing uses which require them, and not released for other uses unless there is a clear surplus of suitable sites in the area, and no realistic prospect of development for that purpose in the foreseeable future.
>
> (ibid.: Para. 12)

Of course, the guidance is careful to stress that once allocated, there should be no switch to other uses. For example, it would be inappropriate to allow retail development within a distribution park, or to allow the creation of a factory outlet from an existing warehouse. Transport has been a key driver of economic development at the rural-urban fringe: it has created opportunities that have been grasped by a range of service industries. But it is also a barrier

5.5 Science park, Ely, Cambridgeshire (Countryside Agency, Photographer Unknown).

to development. A desire to avoid excessive car use has led government to resist the expansion of many fringe sites. However, there is growing pressure for change, especially around the most accessible sites, including those incorporating a parkway function. The last Planning Policy Guidance Note 13 (Transport) called on authorities to make 'maximum use' of accessible sites (DETR, 2001a: Para. 21) including those adjacent to major transport interchanges, known as 'key sites' or 'transport development areas' (RICS, 2000). There is an acceptance that these sites could potentially be made more accessible by public transport, thereby becoming more satisfactory locations for further economic development. This shift heralds a recognition that the rural-urban fringe – or rather key sites in the fringe – have become centres of the new economy, and a departure from the position reported by Shoard (2002: 131–132), that authorities are unwilling to assist such development. Pressures remain to resist a diversification of service sector activity in the rural-urban fringe, because of fears over car use and the impact on High Streets. However, there appears to be growing support for the argument that key sites can become more sustainable, with self-contained semi-urban units perhaps becoming – on a more regular basis – 'edge centres' (if not cities) of the new economy.

Economic survivors in the rural-urban fringe

Farming

The new economy outlined above is a product of rapid transport development in the fringe during the late twentieth century. Farming, on the other hand, is the legacy of a period before this development. Does this mean that agriculture no longer has a place in the rural-urban fringe? Two hundred years ago, farming intensified in the countryside around towns, satisfying the need to feed the growing urban population. There was a locational necessity for agriculture to be close to people: today, this locational necessity has become a potential locational advantage, especially as consumers become more concerned with the environmental consequences of agricultural production, measured in terms or organic versus non-organic cultivation and 'food miles'. Agriculture has been subject to broad political and economic changes in the past 60 years, and it is perhaps difficult to separate general farming from more specific fringe concerns. Concerns over the environmental impacts of intensive farming (since the 1970s) have resulted in reforms of the Common Agricultural Policy (CAP) and a desire to encourage 'agri-environmental' initiatives and practices. In this context, Bickmore Associates (2003) suggest that food production in the fringe has been relegated behind environmental and access objectives. In a sense, farming 'survives' in the fringe, but it does not prosper, often because it comes into conflict with other uses. This same point was made in the late 1970s by the Advisory Council for Agriculture and Horticulture:

> The outer edges of our cities and towns are an acute source of conflict between urban and food production interests. They contain

106 An economic fringe

significant proportions of agricultural land of very high quality with considerable food production potential. At the same time, urban fringe provides for many millions of the local urban populations their chief source of rural recreation and enjoyment. Yet neither the full agricultural nor the full amenity potential of these countryside areas is at present being realised.

(Advisory Council for Agriculture and Horticulture, 1978: 20, 37)

Clearly this comment points to problems in the way different uses and competing demands in the fringe are integrated. Farming in the fringe enjoys potential advantages – including access to nearby urban markets – but it is the problems that receive most attention. Large-scale farming has been pushed out of the fringe: fields are now truncated by roads and railways, making it difficult to move livestock. For the same reason, farms have become increasingly fragmented, with many becoming little more than smallholdings. Bickmore Associates (2003), in a study dealing particularly with the state of farming in the rural-urban fringe, have noted that there is little hard evidence of the problems experienced by farmers though there is anecdotal evidence of labour supply difficulties and farmers having to deal with persistent fly-tipping (Figure 5.6). In some cases, farming and more 'urban' uses have failed to get along, with fly-tipping occurring along main roads and interchanges, especially where illegal dumpers can duck behind warehouses and unload unwanted white goods and mattresses.

5.6 Fly-tipping, Rotherham, South Yorkshire (Countryside Agency, John Morrison).

But there are other cases where agriculture has managed to adapt, integrating with other uses in the fringe, and successfully exploiting locational advantages. In some instances, so-called 'adaptive farms' have developed into commercially successful ventures in the fringe. Bickmore Associates (2003: 27) cite examples in west Lancashire where high-value crops are cultivated close to the urban edge and distributed locally. In Bath, the 'Real Bath Breakfast' accreditation scheme stipulates that hotels (which wish to participate in the scheme) must source breakfast ingredients from within 40 miles of Bath Abbey (ibid.: 35). Similar schemes across the country aim to promote local farming and reduce 'food miles'. Concern over food miles and the subsequent desire among some consumers to switch to local seasonal produce may mean that rural-urban farms can more easily respond to environmental concerns. Certainly, 'farm shops' have been a feature of some fringes for many years. Today, farmers are extending 'direct marketing' initiatives and drawing up contracts with local suppliers: again, proximity to large urban markets is a major advantage when developing such strategies.

In general, farming in the UK has fallen on hard times. In 1998, total income from farming (TIF) fell to just over £2.5 billion, its lowest level for a quarter of a century. Its contribution to the national economy (GNP) fell to below 1 per cent in the following year. The foot and mouth disease crisis of 2000 and the longer running BSE (bovine spongiform encephalopathy) pandemic of the 1990s dented the public's confidence in the sector. This is now reflected in farming's own assessment of its future prospects: in the late 1990s, a tenth of UK farmers believed that farming 'had no future' and 56 per cent had diversified into non-farming enterprises (Countryside Agency, 1998). The same global shifts – reflected in the structure of the national economy – that have fuelled the emergence of a new service-sector economy at the fringe, have also affected farming. As far as the UK is concerned, farming along with primary and extractive industry is likely to decline in importance as food and the products of heavy manufacturing (and aggregates) are sourced from abroad. But globalization creates a certain type of product with particular characteristics: factory farmed and from the other side of the world. A backlash against globalization may strengthen the hand of farming in the fringe as it re-realizes the locational market advantages that it enjoyed two centuries ago, and as it returns to many of the (non-fertilizer, 'organic') practices that disappeared at about the same time.

So this is the first point with regard to agriculture: if farming in the fringe can resolve its conflicts with urban uses, then it may begin to realize the potential advantages of its location (Figure 5.7). The second point is very similar: there is often a bigger market for non-farming enterprises in the fringe. Indeed, in a study of 'farm diversification' in the West Midlands it was found that two-thirds of the diversified farms were located within 5 km of Birmingham and Coventry: 'an urban fringe location, therefore, would initially appear to be an important factor in enhancing the earning power of alternative enterprises. Indeed, 82% of farmers stated that proximity to a major urban market did influence their decision to diversify' (Ilbery, 1991: 213). Nationally, two-thirds of farms now engage in non-farming enterprises. Land next to rivers and

5.7 Farming with airfield beyond (Countryside Agency, Liane Bradbrook).

streams is opened up to fee-paying anglers; land is leased to companies organizing executive team-building weekends (camping, paint-ball, quad-bike racing and so on); farms have been turned to 'horsiculture' with riding events or pony-trekking; old farm buildings may be converted and leased as weekend cottages or business units; farmers may venture into food processing, exploiting the current stampede for everything and anything with the label 'organic' – hence many farms have entered the competitive cheese and yoghurt markets. The list of potential diversification is endless. Some farms – particularly those in the fringe – have taken on a broader community and educational role, opening themselves up to visitors.

But although there is potentially a bigger market for such activities and services in the fringe (see Beauchesne and Bryant, 1999: 327), it is apparent that many farmers still find it difficult to break even. Bickmore Associates cite the example of an unnamed farm manager running a traditional farm now reborn as an unnamed 'community farm' in an unspecified location. The farm had been enveloped by a housing estate at the urban edge and responded by opening its doors to visitors. But income derived from gate receipts was insufficient to meet the costs of feeding animals or making physical alterations to the farm to satisfy health and safety regulations. Perhaps more critically, local people were unable to afford organic produce from the farm (the price of which had to reflect additional production costs) and so this produce tended to be purchased by middle-class people travelling to the farm by car from other nearby towns. But even this income did not meet all the farm's running costs and so this enterprise remains dependent on sponsorship and charitable donations (Bickmore Associates, 2003: 48). Diversification can be costly and may not always rescue farms from the financial doldrums. It may also be opposed by lobbying groups or nearby residents (Lobley et al., 2002) who fear that a move away from simple cultivation may be a precursor to more radical redevelopment. Arguably, farming's economic position in the fringe is fragile because of a failure to resolve conflicts with other uses. Infrastructural development has fragmented farm-holdings, making them less commercially viable; and money is wasted on protecting farmland from illegal tipping and vandalism. But there has also been a failure to understand the benefits of fringe locations, and only limited local efforts to harness these benefits. Arguably, agriculture could do more than simply 'survive' at the rural-urban fringe.

Mineral extraction and waste management

Mineral extraction and commercial waste management are the final category of economic activity examined at the fringe (Figure 5.8). It is certainly misleading to describe them as 'survivors'. While it might be said that extractive industries have declined in recent decades, commercial waste processing and recycling has become big business as a result of rapid urban growth in many areas of the world, and the UK is no exception.

Although some 'soft aggregates' (sand and gravel) may still be sourced in fringe areas, a decline in metal, iron and coal extraction has left a legacy of abandonment in many peri-urban hinterlands; this was examined very briefly in Chapter 3. Old quarries provide opportunities for re-use including for the purpose of waste management, in the form of landfill. But before turning to re-use and to waste, it is perhaps worth reflecting briefly on the quarrying of aggregates for the construction industry, which retains a presence in many areas of fringe. The siting of extractive industries is a controversial issue and mineral extraction is an activity under intense pressure. This pressure results from vociferous public concern over the siting of what is generally considered to be an 'un-neighbourly use' that results in an erosion of landscape quality. Pressure also results from the proliferation of landscape designations and protection measures in both remoter countryside and in the rural-urban fringe (e.g. in terms of green belts). The total area of English countryside covered by landscape, wildlife, and heritage designations continues to increase annually, and each type of designation has its own group of champions and supporters. But even in a more pressured context, the construction industry is dependent on a reliable supply of both soft and hard aggregates: they have to be sourced from somewhere. Given this reality, the fringe becomes the most acceptable location for extractive industry (this same political expediency has resulted in other un-neighbourly uses being directed to the fringe – see Chapter 2). Of course, the location of quarrying activity is not only a function of politics and subsequent planning decisions, but also a consequence of geology. In those areas with the right aggregate opportunities, commercial extraction is likely to be directed to the fringe. This is perhaps an obvious point and one that we do not need to dwell on. It is also the case that mining operations may not be viewed as a significant contributor to the make-up of

110 *An economic fringe*

5.8 Landfill, Ewelme, Oxfordshire (Countryside Agency, R. Pilgrim).

the economic fringe. The Standard Industrial Classification figures supplied by the Office for National Statistics (ONS) lumps mining with commercial energy and water supply and distribution. This grouping accounted for 0.1 per cent of business stock across England in 2003 and had declined by 10.4

per cent from 2000 (Countryside Agency, 2004: 208). Mining is not a big employer in the rural-urban fringe, though its product fuels wider economic growth in the UK more generally. Its influence in the fringe is more physical and aesthetic than economic, and at the current time, closure and abandonment is a bigger issue than the granting of new minerals permissions.

However, it is useful to use this economic activity – despite its limitations – to illustrate how different functions (perhaps economic and communitarian) in the fringe might be integrated. Linking to ex-minerals workings, the ODPM has recently affirmed government's commitment to promoting the role of woodland in the restoration of former industrial land, making the point that disused quarries have an important role to play in the fringe. The case cited links to broader plans for the regeneration of the London and Kent Thames Gateway:

> In the early 1990s, Thames Chase Community Forest began work with the quarry products company, Tarmac, to restore the East London Quarry for recreational use. Footpath links were established to nearby Rainham and to Hornchurch Country Park – another land reclamation scheme developed by the London Borough of Havering on a former landfill site which had previously been RAF Hornchurch. The woodland has now been linked by footpaths and cycle-ways to Upminster Bridge underground station, artists have worked with local people to design seats and signs, the Forestry Commission has added further extensive areas of woodland and the mosaic of 160ha of contrasting green spaces is proving increasingly popular with a wide cross section of the public.
> (ODPM, 2004b: Para. 4.15)

The suggestion, of course, is that it is possible to find new uses and to 'beautify' the fringe's industrial legacy. This issue was discussed at some length in Chapter 3. But active quarries are very different: noise and dust may force other uses to keep their distance, not only recreational uses but also other economic activities. The campus-like office parks described earlier are unlikely to want to locate next to quarries. Trucks laden with hard aggregates or sand thundering past clean white offices might disrupt the 'air of quiet good taste' described by Castells and Hall (1994: 476). At a more general level, this suggests that a range of different economic uses – commercial waste processing, manufacturing as well as quarrying, each of which is potentially noise generating – remain incompatible with other aesthetic or social functions in the fringe. But experience suggests that this is not always the case. For example, the Maidenhead, Windsor and Eton flood alleviation scheme, completed in 2002, and known as the 'Jubilee River', performs not only its primary function of protecting local towns from flooding, but has also created new opportunities for recreation (along 'river' banks); attempted to re-create wildlife habitats, especially wetlands, that had been lost along the Thames; and is linked with minerals extraction operations. Indeed, the scheme has been associated with gravel extraction and processing, as well as the transfer of unprocessed materials (exploiting the 2 million tonnes of minerals extracted during construction). In the early planning stages, the construction of the main channel in gravel-bearing land led to a suspicion that the scheme was simply a back-door means of permitting gravel digging (Clearhill, 1994). This fear, however, has proven to be unfounded.

In this instance, it has been possible to integrate economic uses with other functions though it would be naïve to argue that a happy alliance of industry and recreation is possible in all instances. Landscapes may be inherently 'multifunctional' but there are limits to what can be achieved in a shared space. This is also true in relation to commercial waste management. Waste processing may not be the most profitable business in the fringe, but it is possibly the most critical. The south east of England (excluding London) will generate between 0.92 and 1.2 million tonnes of additional municipal solid waste (MSW) over the next 20 years (to 2026) linked to household growth. The MSW total will rise to between 6.7 and 7.0 million tonnes and MSW comprises only 12 per cent of the region's waste. It has been calculated that up to 35,000 additional homes will need to be provided each year if economic growth in the region (measured in terms of productivity and output) is to be maintained. Growth produces waste; and waste has to be managed if it is not to become a barrier to growth. Put simply, successful waste management is critical to economic prosperity and therefore one of the key functions of the economic fringe.

Landfill still accounts for 85 per cent of municipal solid waste disposal (the other main management approaches are recycling and composting) despite attempts to break away from reliance on landfill from the 1980s onwards. In the south east of England, it has been estimated that 75 small and 15 large new landfill facilities will be required by 2025, alongside almost 800 small and 300 large recycling facilities, and around 250 composting facilities (South East England Regional Assembly, 2005). Sites will have to be found for all these additional facilities. Two key factors will be critical in determining where new facilities are located. First, towns and cities produce by far the largest proportion of municipal, commercial and construction waste. Second, a 'proximity' principle (DETR, 1999: Para. 6c) in waste management dictates that waste should be processed close to where it was generated (to guard against the movement of waste over long distances by road, and prevent key areas from becoming national dumping grounds). These factors point to the fringe as being perhaps the most acceptable location for waste processing. Indeed, Planning Policy Guidance Note 10 (Planning and Waste Management: DETR, 1999) suggests that management facilities should be located in industrial areas, especially those containing other heavy or specialized industrial uses, degraded, contaminated or derelict land; in working and worked out quarries – landfill is commonly used in quarry restoration but there may be opportunities for other types of waste management facilities at some quarried sites; in existing landfill sites – where composting facilities may be conveniently located; in existing or redundant sites or buildings – which could be used, or adapted, to house incineration or materials recycling facilities, or composting operations; in sites previously occupied by other types of waste management facilities; and in other suitable sites located close to railways or water transport wharves, or major junctions in the road network (ibid.: Para A51). All of these sites and characteristics – the redundant buildings, the old quarries, and the major road junctions – are found in the fringe. In particular, the legacy of old minerals workings and the network of major road intersections make the fringe seem like the ideal location for waste processing. The proximity principle together with the aim of ensuring 'regional self-sufficiency' ('most waste should be treated or disposed of within the region in which it is produced', ibid.: Para. 6b) reinforce this view. More than 1,300 new waste processing facilities of various sizes will need to be provided in the south east of England over the next two decades. At the beginning of this chapter – and also in Chapter 1 – we pointed to the fringe's role in serving nearby towns and cities. Waste processing is a crucial environmental and economic service: a cornerstone of future economic growth. Indeed, all such service functions – waste management, water supply and treatment, and energy generation – are essential prerequisites to growth: they are all part of the economic fringe in the sense that they serve a broader economic purpose. This differentiates them from the new economic life of the fringe, examined in the first half of this chapter, but it illustrates that the fringe performs a vital economic role. It is not a dead landscape where nothing happens. It is an arena for a new economy, and performs functions essential for continued regional prosperity.

Conclusions

We have painted a rather positive picture of the fringe in this chapter: as a new economic space, able to out-compete nearby towns and cites; and as the building block for regional economic growth. This picture might seem rather fanciful, but the conditions have been created through physical development and through policy to assign a distinct economic role to the rural-urban fringe. The expansion of towns and demographic decentralization beyond fringe belts has resulted in a densification of peri-urban road networks. This trend looks set to continue and therefore more opportunities will arise – more by fortune than by design – for new economic hubs in the fringe. The hubs that already exist are diversifying; and they are gradually becoming keyed into transport networks that are not entirely car-based. The parkway stations are an example in hand. Some may become edge-centres of the service economy incorporating a mix of retail, office and leisure functions. As they evolve, pressure may grow to permit housing development (as in the case of Bluewater in Kent), bringing such centres closer to Garreau's notion of 'edge cities'. This may not bode well for established cores and government remains committed to ensuring that there is greater differentiation between the economic functions of new hubs and existing centres. There is clear scope for such differentiation: out-of-town activities often need different things and government has acknowledged that activities such as distribution and bulk retail benefit from a fringe location. At the same time, service functions and especially waste processing are also more suited to the fringe. Although it is difficult to envisage landfill and retail wanting to be located at the same transport hub, the latest waste management facilities (perhaps engaged in specialist recycling) may sit more comfortably in the 'impeccable landscaping' of the modern science park.

Perhaps the key message emerging from this chapter is that the fringe has a diverse economy, comprising a mix of new service functions, the prerequisites of wider economic prosperity, and the 'economic survivors' who occupy a more precarious and uncertain position. In terms of the last group, we have focused only on farming in this chapter but might also have looked at commercial forestry or various recreational uses that have not always enjoyed great economic success. The key issue for farming at the fringe seems to be the sector's past inability to reconnect directly with the urban market. Its fringe location is often viewed more as a burden than an opportunity, with concerns over illegal dumping, subsequent land degradation and land fragmentation resulting from road building, receiving most attention. It appears that farming at the fringe needs clearer guidance and more of a helping hand from national and regional government. Of course, some farmers no longer wish to farm: they view illegal fly-tipping as an excuse for them to disengage from agriculture and live in the hope that the degradation of their land will encourage planning authorities to grant permissions for alternative, more profitable uses. But farming is already a potentially profitable use in the fringe if public awareness of organic production and the need to reduce the 'air miles' burden, can be harnessed and people encouraged to buy local produce. However, there are admittedly far more profitable economic uses in the fringe. We have emphasized in this chapter the economic dynamism of the rural-urban fringe, expressed most clearly in the post-war emergence of a new economy, exploiting the fringe's locational, space and cost opportunities. This economy has now become the defining economic – and also aesthetic – feature of the rural-urban fringe along with the arterial and orbital road networks that made its emergence possible.

6

A sociocultural fringe

Introduction

In Chapter 2, we described the fringe as a 'landscape without community', and this is an image of the rural-urban fringe supported, at least in part, by Daniels (1999) and Shoard (2002). But this view seems to have been contradicted in the last chapter. Retail and leisure-based 'edge centres' are sometimes associated with residential development and particularly with red-brick 'estates' organized on culs-de-sac. It is also the case that in the years before green belts (see Chapter 8), some expansion of towns occurred along arterial roads to form low-density 'ribbon development'. But the characteristics of the fringe – a functional landscape of unimproved grassland truncated by major roads and dotted with low-density warehousing and essential services – can make it an inhospitable space for living. However, the notion of a 'sociocultural' landscape should not be conflated or confused with the idea of a 'residential' landscape. The fringe may, in large part, be a landscape without community, but it does have a community of users: people with various objectives in mind who experience the fringe (or might experience it in the future) for informal recreation, for formal leisure, or for employment. This chapter examines three major issues, or dimensions, within the sociocultural fringe. Because of the reality of some housing growth in the fringe – particularly around transport hubs including parkway stations (see Chapter 5) – we begin by focusing on the quality of housing development in the fringe. But the major focus of this chapter is how the fringe is used for non-residential and non-employment uses: how people experience the fringe for recreation, for education or for health reasons. The benefits of having open access to the fringe for these purposes has been highlighted in the government's recently issued PPS 7 (see Chapter 8). The second objective is to focus on the fringe as a space for recreation and experience. This leads into a third and final objective: to consider how accessible the fringe is for potential users. Accessing major hubs, where a collection of economic and service uses may be concentrated, is a fairly straightforward task as we saw in the last chapter. But how connected are towns and cities with the unimproved grasslands and open spaces of the rural-urban fringe?

How communities experience the fringe – and how they access it – have been highlighted as major areas of concern for future policy (Countryside Agency, 2004, 2005). The emergence of a new economy on the fringe, and the success of this economy, may result in areas becoming inaccessible. It may be politically expedient to locate low-density development, quarrying or less attractive service functions around major transport hubs, but the concentration of these activities in the fringe may close off areas to the general public. And because there is no 'political voice' representing the fringe (see Chapter 2), such uses may swamp particular areas creating no-go zones in areas where people previously wanted to go. Issues of access vary in different parts of the fringe. At the inner urban edge, allotments will be accessible to those who wish to become part-time smallholders, though the allotments themselves may present a barrier to accessing the fringe beyond. Further into the fringe, some transport hubs might provide access nodes to woodland or unimproved grassland, but proximity to busy roads may reduce the amenity of such spaces. It is the woodland, open spaces, and waterways in between major roads and railways that people may want to access for recreation and these are the areas that are often least accessible to urban communities. In some instances, it is the relationship with other fringe functions that limits the utility of the fringe's sociocultural 'asset': bypasses may present a physical barrier blocking access to open spaces beyond; fly-tipping may degrade the landscape, creating contamination and hazards that pose a risk to health; or land-use planning may seek to compartmentalize uses, closing areas to public access.

Much has been said, in policy and in a wider literature, on the potential of open countryside around urban centres. There is an expectation that people will wish to commune with nature, and reconnect with the natural environment. Others will simply want to walk the dog. There is much talk of 'a re-engagement' between town and country, with city-folk flocking to the fringe to learn more about the countryside, the wildlife, and the habitats: education will create a new awareness and new appreciation of rural issues. And the cultural schism between rural and urban will be bridged. However, the reality is that the fringe has an entirely different sociocultural function for many visitors. Rather than communing with nature and learning to recognize the mating call of the *Acrocephalus Palustris* (i.e. the Marsh Warbler), they choose instead to drive to an out-of-town leisure complex for tenpin bowling, to catch the latest Hollywood release, to dine with friends, or to spend an hour at the gym on the way home from work. These are as much aspects of the socio-cultural life of the fringe as the enjoyment of open countryside, though they may be derided by some commentators. Shoard (2002) poses a number of questions relating to the recreation utility of the fringe: 'do people who work in the interface in retail complexes and business parks not need a park for a stroll at lunchtime? Where, for that matter, are they to eat their lunch? Kick a football with their friends? Or watch their children play?' (ibid.: 131). The reality is probably that the business park has its own area of open space, and

Introduction 117

that many of those employed in retail complexes eat at Starbucks. And few will bring their children to work, or want to kick a ball round next to the municipal dump. The complexes derided by Shoard and others have much to offer in terms of recreation: they were designed with this function in mind. But our purpose here is not to pit the recreational tastes of one group against another, or to claim that spending an hour at a video arcade is a more worthwhile way to pass time than playing golf or rambling across a country park. The point is that the new economy of the fringe – examined in Chapter 5 – has its own associated sociocultural function and that many people, who may never visit woodland in the fringe, may nevertheless experience the fringe in a sociocultural sense. The fringe may offer a range of potential experiences and opportunities, and it is the less formal ones which are examined most closely in this chapter. In particular, special reference is made to country parks, woodlands, common land, golf courses, rivers, canals, lakes and reservoirs: all with clear, albeit fairly conventional and middle-class, recreational potential (Figure 6.1). The ways in which these opportunities may be threatened by an expansion of other uses within the fringe is also considered in this chapter alongside the possibility of better integrating the idea of a sociocultural fringe with the fringe's other functions. In relation to landscape history (Chapter 3) or ecology (Chapter 7) such integration might appear natural; but the integration of 'informal' or ecology-based recreation with new economic uses might be less readily achievable. But perhaps the most difficult question that needs to be addressed relates to the 'aesthetic fringe', and the way recreational potential is influenced by 'landscape improvement' or the creation of parkland versus retention of land in an unimproved state.

6.1 Golf course, Alnmouth, Northumberland (Countryside Agency, Andrew Hayward).

118 *A sociocultural fringe*

Before focusing on recreational experiences of the fringe, it is our intention to build on the brief discussion of emergent housing development around transport hubs presented in the last chapter, and consider the experience of 'living in the fringe', which is an experience apparently tied to the emergent service function of many transport hubs: it is also a car-dependent experience and, in many instances, an experience associated with low-quality housing development.

Housing quality and residential experience at the edge

In the last chapter, we reflected briefly on Garreau's (1991) notion of 'edge cities' centred on major transport hubs, and incorporating economic, leisure and residential functions. It was noted that the North American experience of such places is not replicated in the UK though some housing developments have sprung up alongside retail and business complexes in the fringe. The case of Bluewater in Kent was used to illustrate this point, and so too was the emergence of 'commuter estates' next to motorway junctions in the north west of England. The difficulties of restructuring failing housing markets in the north are, it is claimed, being amplified by a proliferation of new house-building in the rural-urban fringe aimed at the commuter market. In some instances, clusters of homes have been permitted by planning committees on green-fields next to new leisure or retail complexes. These are invariably red-brick developments undertaken by volume builders using the same pattern-book designs that are applied across the country. The developments appear formulaic: a mix of one hundred or so detached and semi-detached houses and bungalows, all with pitched-roofed garages; front and rear gardens; double or triple glazing; arranged on a network of closes and culs-de-sac; and each with two or three family estates and hatchbacks in the driveway. The development may be connected directly to a roundabout by its own access road; its residents may use the adjacent service station as a convenience store; outgoing mail is collected once a day from a postbox on the main road; the Ikea superstore, just a stone's throw away, supplies all the estate's furnishing needs; and it's at least a mile along a tarmacked walkway to the first road sign announcing the edge of the nearest town.

For some, this is a picture of the ultimate in convenience living. Many people aspire to spacious new homes, and these tend to be more affordable when located in developments of the type described above. The need to travel, by car, to work or to school has become ingrained in the daily pattern of life for many people in the developed world: living in these fringe developments may be seen as an advantage as peri-urban roads are likely to be newer, wider, and faster. But for others, the lifestyle offered by out-of-town estates is a car-based, soulless, modern anathema. To be discarded in the midst of an anonymous landscape of major carriageways, concrete flyovers, electricity pylons and warehouses: is a worse fate possible? From a planning perspective, central and local policy makers would probably agree with the latter view: although many existing centres are facing pressure to grow – and some cores are proving unattractive to new home-buyers – the prevailing view is that policy

6.2 Housing, allotments and canal, Nuneaton, West Midlands (Countryside Agency, Julia Bayne).

should steer development away from out-of-town hubs and focus instead on the much-lauded 'urban renaissance' or the creation of 'sustainable extensions' that integrate communities with existing urban infrastructure rather than leaving them out on a limb (Figure 6.2).

In 1998, the Urban Task Force (UTF) under the chairmanship of Richard Rogers was handed the task of advising government on the best options for achieving sustainable growth in England (in the context of official growth projections and a realization that many urban areas had suffered decades of neglect). The Task Force was primarily concerned with the revitalization of inner city areas, advocating densification and compaction of existing cores when it eventually reported in 1999. It emphasized the need to focus on inner areas, and to consider development options outside existing boundaries only when the capacity for land-recycling had been exhausted. This concern for 'sustainable urban form' and for the creation of more compact cities has been a defining feature of thinking on planning and urban development since the 1990s (Williams et al., 2000; Jenks and Burgess, 2000). But pressure on existing cores, spilling over into the fringe, is likely to lead to at least some calls for urban extension or perhaps – in fewer instances – an acceptance of some limited 'nodal development' of the kind described above. Though it is not a case of 'resistance being futile', there is perhaps a need to consider the best way to plan for and manage such development in the rural-urban fringe. The 2003 'Communities Plan' (ODPM, 2003a) continues to warn against excessive development of green-field sites for new housing or economic use; however, it notes that opportunities to development brown fields can exist 'in and around' urban areas (ibid.: 5). We noted in Chapters 1 and 3 the opportunities to recycle former airfields or ex-military sites within the fringe, though these may not always be well served by modern road and rail infrastructure: in some instances, airfields may have been enveloped by residential development and might, therefore, provide apparently ideal sites for extension or infill development. Those that remain 'stand-alone' may be less easy to integrate with existing urban infrastructure. However, there may be some opportunities to recycle ex-industrial land close to transport hubs (especially at parkways, which may also be served by local buses). In these instances, an opportunity exists to create the kinds of hub estates described earlier or maybe to develop something a little better.

This is not to suggest that hub development is ideal, but some at least half-contained 'edge centres' where there is a mix of employment and recreational opportunities may serve the needs and preferences of some households. In North America, Sullivan (1996) has suggested that 'cluster housing' in the fringe may be preferable to more suburban development, arguing that suburbs 'often destroy rural character. Planners have used standard design requirements built into conventional suburban zoning and sub-division ordinances to guide development at the fringe' (ibid.: 292). As an antidote to this problem, the development of 'closely-knit villages surrounded by open space' is advocated (ibid.: 293), which could be said to accord with the idea of 'eco-villages' within a 'new geography' of 'mostly . . . small and highly self-sufficient local economies, in which most of the things we need were produced within a kilometre or two of where they were used' (Trainer, 1998: 83). It would be fanciful to suggest that the hub developments bolted onto motorway junctions around many English towns come close to the ideal of well-designed 'cluster housing' or ecologically sound and self-sufficient 'eco-villages'. They clearly fall well short of these ideals. Likewise, the developments display no concern or appreciation whatsoever for lightening the ecological footprint or achieving greater sustainability. Some are associated with other nearby uses, but there is no hard evidence to suggest any real interaction or integration between these different functions. We just assume that residents use the nearby multiplex. Existing hub development is merely opportunist, erected quickly and cheaply for maximum profit: the only real issue of interest relates to the mysteries of a political process that would permit such development. Perhaps it is merely further evidence of the low esteem in which the rural-urban fringe is held: because the landscape is attributed with so little intrinsic quality, no-one believes that poor housing will make any difference.

But it is likely that hub developments will be the exception rather than the rule in the future: those that are granted permission will be few and far between; there will be an expectation that some connection can be made between housing and the fringe's new economy, and between housing and nearby services. They are also more likely to be located on ex-industrial sites that are deemed suited to residential development, because of potential integration with public transport links. And, there is likely to be greater concern for design, for context and for 'ecological modernization'. If a case can be built to value the rural-urban fringe, then it becomes more likely that design issues will be prioritized. Of course, few such developments are likely to be permitted within the green belt (see Chapter 8), though in some instances, local management frameworks for the fringe might look more favourably on creating new communities in the fringe that could demonstrate more sophisticated design standards and a greater concern for local ecology. Although the purpose here is not to sell any particular manifesto, it might be appropriate in some instances to aspire to the development of 'greener' housing in the fringe, especially if it is the purpose of future fringe policy to create stronger linkages between town and country, perhaps through a demonstration of how human influence in the landscape might be softened. Congreve (2003) has lamented the current lack of concern for environmental improvement in modern house building, and the construction industry's failure to engage with 'ecological modernisation' (ibid.: 329). Perhaps opportunities exist to bring 'real communities' into the rural-urban fringe, and to create a different kind of sociocultural function that seeks to engage with environmental debate.

It must be possible to improve upon the horrendous hub estates that have emerged in some fringes. There are examples, across the UK and not only in the fringe, where traditional forms of housing and residential experience have been challenged, and these perhaps offer some pointers to the development of 'cluster housing' in the fringe. The BedZED (Beddington Zero Energy) development at Hackbridge in South London, for example, is often held up as the kind of housing scheme that might be replicated elsewhere (Figure 6.3). The development is part of the Peabody Trust's portfolio of recent development schemes in London: 'it aims to be a beacon, to show how we can meet the demand for housing without destroying the countryside. It shows that an eco-friendly lifestyle can be easy, affordable and attractive: something that people will want to do' (http://www.bedzed. org.uk/about.htm; accessed 9 June 2005). The scheme – which comprises 83 units for sale or rent – has a number of characteristics that set it apart from other developments:

- It utilizes, where possible, building materials selected from natural, renewable or recycled sources and wherever possible brought from within a 35-mile radius of the site;
- It also has a combined heat and power unit able to produce all the development's heat and electricity from tree waste (which would otherwise go to landfill);
- The homes themselves have an energy-efficient design: houses are south-facing to make the most of the heat from the sun, and they have what is described as 'excellent insulation and triple-glazed windows';
- A water strategy is also in place, able to cut mains consumption by a third: the developers installed water saving appliances that make 'the most of rain and recycled water';
- In order to promote greater connectivity to its surroundings, the site has a green transport plan that aims to reduce reliance on the car by cutting the need for travel (e.g. through Internet shopping links and on-site facilities) and providing alternatives to driving such as a car pool;
- Finally, at a very basic level, there are recycling bins in every home.

122   *A sociocultural fringe*

6.3 BedZED, Hackbridge, South London (Bill Dunster Architects).

Nowadays, no analysis of sustainable living or ecological housing is complete without some mention of BedZED: a simple Internet search on the name generates close to 16,000 hits; its 'innovative design' and 'environmental credentials' have been lauded worldwide. The World Wildlife Fund (WWF) claims that BedZED makes 'sustainable living easy, attractive and affordable' (http://www.wwf.org.uk/wssd/briefings/homes.pdf; accessed 9 June 2005) adding that:

> BedZED has significant benefits for the environment and the local community. A village square with a nursery and café provide a focus for the mixed development that integrates home and the workplace. On-site water treatment, rainwater collection for sewerage, an integrated transport provision, energy-efficient design (reducing space heating needs by 90 per cent), and a renewable energy supply, mean each home's environmental footprint is radically reduced. ZED buildings are also healthy, being allergy free and having excellent ventilation. Materials used are from local, recycled or certified sources. Furthermore, the adjacent landfill is to become a working parkland with urban forestry, and space for organic produce, energy crops and wildlife habitat. Bulk delivery of groceries and an electric car share scheme also reduce the environmental impact of transport associated with the development.

BedZED does have a small, but growing, number of critics. Davis (2004), for example, has pointed out that a 'lack of attention to streetscape and public space . . . has rendered BedZED rather sterile' and lacking social cohesion (ibid.: 3). But it clearly represents a potential improvement on the schemes described earlier: and it is of a size that could be suited to the fringe, providing a possible blueprint for 'cluster housing'. The scheme also has the advantage of comprising 'live–work' units: but living and working *on the edge* of a built-up area, and enjoying the benefits of proximity to local services is one thing; feeling isolated *within the fringe* and disconnected, by distance, from existing centres may not appeal to many people.

However, if hub development is to occur within the rural-urban fringe or if 'cluster housing' becomes more desirable than a concentration of dwellings in urban extensions or suburban sprawl, then the need to identify development models suited to the fringe will increase. There is perhaps greater pressure in the fringe and in the wider countryside to ensure that developments 'blend in' rather 'stand out' within their context, in an ecological as well as an aesthetic sense. This is something that the Hockerton Energy Project (near Newark in Lincolnshire) tries to achieve (Figure 6.4). The scheme comprises 'a parallel row of south facing, earth sheltered terraced houses' (B. Edwards, 2000: 139), which integrate a number of low energy and ecological considerations: 'water, for instance, is locally gathered, cleaned and used as grey water for certain domestic functions' (ibid.: 139). The Hockerton project is specifically designed to be a 'demonstration scheme' and as such represents an aspiration more than a model that might be replicated elsewhere. It is an experiment in sustainable housing design that tries to combine residential use with energy production and various forms of recycling. Its key targets include: 90 per cent energy saving compared with a conventional house; self-sufficiency in water; EU bathing water standard of sewerage effluent from site; three-month temperature time lag due to earth sheltering; 70 per cent heat recovery from extracted warmed air; one fossil-fuelled car per household; 100 per

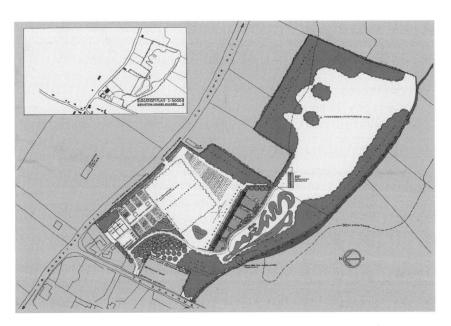

6.4 Sketch showing Hockerton Energy Village in its 'ecological setting' (Brenda and Robert Vale).

cent self-sufficiency in fruit, vegetables and dairy products using organic/permaculture principles; eight hours per person per week in support activities. The scheme has a far lower media profile than BedZED and perhaps represents a more extreme form of 'green housing', with residents having to commit to a particular lifestyle and 'encumbered by requirements to on-site productivity (creating jobs on-site) and to running an educational outreach programme' (www.coldhamarchitects.com; accessed 09 June 2005). But if Hockerton is at one extreme, then many current developments in the fringe are at an opposite extreme, representing the antithesis of green living and usually failing to integrate with the fringe's other functions.

The intention here is simply to illustrate the huge divide between what is possible and what is actually happening within the fringe. The BedZED and Hockerton schemes are a far cry from the type of low-density red-brick, low-quality, energy-inefficient housing, that currently litters the English landscape. Even if Hockerton cannot be mainstreamed, there is now increased support for moving towards improvements in design and environmental standards. These concerns are raised in general terms in Planning Policy Statements 1 and 7 (see Chapter 8) and are likely to feature prominently in a future PPS 3 (replacing the current planning guidance on housing in England). In 2003, the Chartered Institute of Housing issued a joint statement with the Royal Town Planning Institute, arguing for a greater promotion of design and environmental standards by planning authorities: something that many have apparently failed to do in the past (Congreve, 2003). B. Edwards (2000: 140) suggests that future housing should fulfil five basic conditions. It should be 'low resource using' in terms of energy, water and land consumption; it should be 'safe', promoting security through design; it should be 'healthy', contributing to physical and mental well-being; it should be 'productive' in a social sense (promoting community cohesion) and connected to jobs either through live–work arrangements or because of proximity to employment; and it should be 'beautiful' aesthetically and ecologically. These are very broad, but also very basic, conditions.

For a variety of reasons, the rural-urban fringe is a 'landscape without community' (see Chapter 1). First, housing is not a major land use in the fringe; areas of extension or suburban development have ceased to be fringe in the sense that we use in this book. Second, planning restrictions and a presumption against development in 'open countryside' limit the scope for residential development. Third, where development does occur – centred on transport hubs – its form tends not to lend itself to community building. These are dormitory estates whose residents are disconnected with the landscape around them. The sociocultural experience of living in the fringe may not be wholly positive. And yet, if economic development and employment uses – distribution, retail, business and office parks – are appropriate within the fringe around transport hubs (see Chapter 5), then pressure may follow to introduce more housing into these areas for the sake of the 'homes–jobs' balance. This has already happened at Bluewater in Kent and at countless motorway junctions up and down the country. Perhaps the biggest problem (besides the drawbacks described earlier) with these developments is that new residents are not attracted by proximity to employment uses, but by access to roads. Their aim is not to take up nearby employment – helping local authorities achieve a better homes–jobs balance – but to secure a convenient base from which to commute. This is particularly important for dual-income households where partners travel to work in different directions. Where one partner works in London and the other in Birmingham – opposite ends of the M40 – the 'best' residential location might be close to a motorway junction at Oxford or Banbury. If, on the other hand, the two workplaces are in Manchester and Liverpool, then somewhere on the M62 north of Warrington would be ideal. Thus the fringe performs a very limited residential function, though this might be extended through 'green initiatives' of the type examined earlier. However, hub development has an entirely different rationale from these alternative initiatives. It is not about environmentalism or sustainability; it is merely about access. Green initiatives in the fringe cannot replace hub development, but they might be able to demonstrate the current rift in building standards between the good, the bad and the downright ugly. Cluster housing offers an alternative vision for living in the fringe, but one that is a long way from being achieved.

Recreational experiences at the edge

The fringe has a 'community of users' (or potential users). At the beginning of this chapter, we suggested that it has a sociocultural function not only in terms of 'greener' and more environmental pursuits, but also in terms of basic urban opportunities of the type found in leisure complexes and retail centres. It also has other 'special' opportunities. In Chapter 1, we noted the concentration of old airfields in the fringe. Many of these are now associated with motor sport, Sunday markets and car-boot sales; some accommodate leisure flying, aviation schools, or visitor centres. A few host festivals and fêtes, drawing people out onto the fringe for rock concerts. Though not strictly within the rural-urban fringe, Long Marston Airfield near Stratford-upon-Avon – once famed for the Phoenix Festival – played host to the 'Godskitchen Global Gathering' in July 2005. Also linked to flying are the opportunities that exist to spend an afternoon 'plane spotting'. Terminal 2 at Heathrow Airport, for instance, has a dedicated viewing area on its roof, with a cafeteria providing refreshments. And what better way to round off a day of observing 'airport activity' than by visiting the Heathrow Visitor Centre to see an 'interactive exhibition covering the airport's past, present and future'. Though rock festivals, car-boot sales and plane spotting may not appeal to everyone, they are serious pursuits for many users of the fringe. So too are visits to multiplex cinemas or the bowling alleys mentioned earlier: the fringe is the arena for a diverse range of sociocultural experiences. But for those agencies and writers concerned with open space, with the environment and with educational experiences, it is 'informal recreation' at the fringe that receives most attention (Figure 6.5).

The fringe is seen as an arena for informal recreation for two reasons: first, its abundance of open space (often comprising unimproved grassland); and

126 *A sociocultural fringe*

6.5 Recreation, canals and a potential 'greenway', Nuneaton, West Midlands (Countryside Agency, Julia Bayne).

second, its proximity to urban areas. In 1999, the Countryside Commission (now the Countryside Agency, but soon to be part of Natural England) announced that the fringe could become 'a patchwork of places where people learn more about their environment and can be involved in its care, creating a sense of pleasure, pride, respect and commitment' (Countryside Commission, 1999: 3). However, it is impossible to quantify how much open land exists in the fringe, and how much is suitable for 'open-air recreation'. The last survey of open areas in London's green belt was undertaken in 1960 and found that 3.6 per cent of land might be suited to this purpose (Thomas, 1970, cited in Shoard, 1979: 89). However, there is a likelihood that spaces considered suited to recreation nearly half a century ago may well be truncated by main roads and punctuated by new economic uses today (irrespective of the green belt designation). There is no way of knowing what spaces are actually available: and it is also probable that what many consider suitable for recreation today (e.g. oases of ecological diversity surrounding derelict buildings) may have been considered unsuitable in 1960 when a more conventional view of recreation prevailed.

Potential recreational space in the fringe tends to come under one of three main categories: first, there are privately owned structures that provide recreational opportunities such as golf courses, driving ranges, football or athletic stadiums and other sports facilities, available often on a fee-paying basis. Secondly, there are publicly provided facilities, such as country parks and reservoirs; and thirdly, facilities not necessarily provided with public recreation primarily in mind but that the general public can nonetheless use. This final category is potentially the most interesting and diverse. It may include

Table 6.1 Recreational land in the fringe

Country estates
Country estates – areas of grassland dotted with trees and frequently associated with lakes and historic buildings – often occur in the fringe as part of landed estates. These were often situated on the edges of towns, to provide easy access to town in the days before roads were in place, but at the same time to provide some measure of seclusion. Examples include Luton Hoo, Windsor Park, Tatton Park on the edge of Stockport, Haddon Hall outside Bakewell, Euston Hall on the edge of Thetford, Broadlands on the edge of Romsey. Parkland landscape is potentially extremely valuable for recreation of various types, including walking, picnicking, playing games, nature study, swimming and paddling. Privately owned country estates which do allow people in, like Broadlands, often charge a fee for access to the mansion and a limited area, including a little parkland, close by. However, they do not usually permit people to wander freely over the rest of the estate, and indeed other parts of the estate which might be attractive for recreation such as woods and riverbanks.

Rivers and canals
Rivers have the potential to play a greater recreational role in the fringe, for walking, swimming, paddling, boating and nature study. However, the extent to which they do this is extremely limited. Public access to riverbanks for non-tidal rivers is extremely limited, as the bed and banks belong to the owners of the land on either side. Owners frequently sub-let navigation and fishing rights, limiting public access. The tolerated access people enjoy along the coast rarely exists alongside rivers and streams unless they are passing through unfenced upland moor and mountain (Shoard, 1999: 89). In contrast to rivers, canal banks are largely accessible,

Common land and woodland
Commons represent some of the most important existing recreation areas in the fringe, notably Epping Forest (see Chapter 2), Berkhamsted Common and Ilkley Moor. Other important urban commons were once situated in the fringe, such as Hampstead Heath, Wimbledon Common, the Strays of York and Port Meadow in Oxford (Shoard, 1997: 339). The provision of an automatic right of public access to all common land in England through the Countryside and Rights of Way Act 2000 may mean that the potential of this type of feature will be consolidated and increased in the future. For many people, a walk in the woods is the archetypal rural experience. Public access to many woods was lost through privatization of the Forestry Commission's estate, which took place during the 1980s and 1990s. Figures on access to woods in the fringe are not available (Shoard, 1999: 58–9). But some global county-wide figures suggest that, like country estates, woodlands and community forests in the fringe could offer a greater range of recreational opportunities.

Lakes and reservoirs
Lakes feature in some country parklands. However, many country parklands remain closed to the public, barring access to lakes situated within them. Since reservoirs serve mainly urban populations, they frequently occur in the fringe. Sometimes the public enjoys access to them for a range of recreation pursuits. Thus at Ecup, in the fringe north of Leeds, public access to the edge exists by virtue of the existence of public footpaths and a bridleway. However, at Hanningfield Reservoir, south of Chelmsford in Essex, public access was found in 1998 to be restricted to views from a public road and a small permissive picnic site

Table 6.1 continued

at least to walkers. Some are accessible to cyclists, but none to riders of horses for reasons of safety. The water itself provides opportunities for boating and fishing. Recreational opportunities provided by canals are concentrated in the West Midlands and South Yorkshire.	on the 12-mile circumference (Shoard, 1999: 94). The Essex Water Company had blocked an attempt by Essex County Council to secure public path status to an existing track down to and along the water's edge. Flooded gravel pits are also a frequent feature of the fringe, since, particularly in south-east England, many towns have grown up in river valleys with gravel deposits, such as the Stour Valley in Kent and the Kennet, Loddon and Thames Valleys near Reading.

canals with towpaths accessible for cycling and walking; old tram lines now used by ramblers; common land on which visitors can hold picnics; and airfields on which a range of leisure activities may concentrate. The category may also include areas of derelict land where illegal trespass is often tolerated by the owners and where people can walk their dogs, fly kites or ride mountain bikes. There is an inherent tension in some fringes between the more 'formal' opportunities in the first and second categories (see Table 6.1), and those less formal activities in the last group. Golf courses and leisure centres bring in a lot of revenue for their owners – and can be integrated with retail and other more formal uses – but their development inevitably deprives less formal activities of space. And while golf and gyms sometimes charge considerable membership fees, free informal recreation is available to a wider spectrum of social groups. Indeed, a formalization of recreation at the fringe is likely to limit sociocultural experience, both in terms of the type of experiences available, and in terms of access. This point returns us to the issue of how much planning and 'sanitization' is needed at the urban edge. The development of golf courses, boating lakes and landscaped leisure complexes might be favoured by those who would prefer to see a neater and cleaner fringe. One can almost imagine the designation of a 'leisure' or a 'recreational' fringe on the latest Proposals Map pinned to the wall of the nearest planning office. But sanitization has a social cost, especially if space becomes privatized. The unimproved open spaces of the fringe provide broader access: if ownership changes hands and improvements are made, then an imperative is created to start generating income from the land. Providing facilities that can charge for access is an obvious way forward.

But there are other reasons why it might be preferable to limit the formalization of recreation and stick with unimproved landscapes. Formal leisure may conflict with other land uses and with the other functions of the fringe. The development of golf courses, for example, may result in the consumption of many hectares of farmland; they may be visually obtrusive and turn the fringe into a theme park of neatly mown grass. Informal recreation, on the other hand, may sit more comfortably with other functions. Land open to public access – under the Countryside and Rights of Way Act 2000 – may still be farmed; it may continue to contribute to energy production in the form of bio-fuels; and it will be available to perform an educational function. Indeed, education can be seen as a major component of the sociocultural fringe. Though few children are observed dodging golf balls and chasing butterflies over fairways, or dredging water hazards for frogspawn, these are the types of activity that are often deemed beneficial in the fringe. There is broad consensus that the fringe offers opportunities for environmental education and hands-on learning, providing children and adults with experiences and skills that they cannot derive from a formal classroom setting (see for example, Taylor et al., 2001; BHF National Centre and Loughborough University, 2004; RSKENSR, 2003). More specifically:

> Children have an entitlement to outdoor play, and there's a growing understanding that this shouldn't just be about rubber matting and bikes and outdoor play equipment . . . of course, children through the ages have lit fires, whittled sticks, climbed trees and made dens. And – although no one ever talked about it in this way – they also, as they did so, absorbed knowledge, took risks, solved problems, kept fit and learned to co-operate with their friends. But today's children are different. Few get to explore the countryside; many hardly play outside their homes. They might have classroom lessons about the Brazilian rainforest, but they've never watched an earwig, splashed through mud, or burnt toast over a smoky fire.
>
> (Wilce, 2004)

In the same way that the fringe has a lawless appeal for the villains of numerous Hollywood movies (see Chapter 4), it may also have a wildness that appeals to children whose homes about the fringe (Figure 6.6). In the nomenclature of planning and public policy, these comprise another of the fringe's 'user groups'. But providing space for whittling sticks and making dens is not a high priority in planning departments. This again leads to the question of how much planning is really needed; if planning cannot accommodate these types of experience then does it have a role in the sociocultural fringe? As far as play is concerned more generally, local authorities today tend to provide penned-off play areas with cushioned floors, no sharp edges, and signs instructing competent adults to 'supervise children at all times'. But do such environments stimulate learning? They probably do to some extent, but do not provide the experience of independence, risk and experimentation described by Wilce (2004).

The UK government agrees that 'fresh air and exercise' can provide a stimulus for learning, not only for children, but also for adults. Its 'liveability agenda' (government's drive for 'cleaner, greener, and safer communities' launched by the Minister of State for Rural Affairs in June 2004) has become a central plank in a raft of new policy initiatives. In 'Greening the Gateway' ('government's vision for the landscape of Thames Gateway'; ODPM, 2004b: Para 1.1), for instance, it is explained that:

130   A sociocultural fringe

6.6 The fringe can have a lawless appeal, Smeatharpe, Devon (Countryside Agency, Pauline Rook).

> ... the local landscape has a valuable role to play in school education and we will encourage educational charities, local authorities, ranger services and others to promote even greater use of parks, nature reserves and other green spaces as outdoor classrooms, studios and laboratories ... there is also a growing recognition that people with practical skills are in increasingly short supply and that there is a considerable training need in the field of horticulture, landscape construction, play leadership, etc.
>
> (ODPM, 2004b: Paragraphs 4.37 and 4.38)

The educational benefits of access to green space are thought to go hand-in-hand with potential health benefits, and numerous studies undertaken over the past few years have sought to highlight the latter. Work by Entec UK, for example, has examined the contribution that woodlands and urban green spaces can make to health improvement, particularly for target groups at risk from such medical conditions as coronary heart disease and obesity (Entec UK, 2003a). Penn Associates (2002) has looked at the work of the National Community Forest Partnership – an initiative launched in 1989 and now comprising twelve 'community forest partnerships' working with 58 English local authorities, with the broad objective of environmental regeneration – and at illustrations of how more accessible and attractive rural-urban fringes may yield physical and mental health benefits. For instance, the Pennyhooks Farm Education Centre in the Great Western Community Forest (surrounding the town of Swindon) tries to meet some of the needs of autistic children. The

Great Western team has worked with a local farmer to obtain funding for classroom facilities that allow it to devise a special needs education programme for the children. This is held up by the consultant as one example of more active promotion of health care and education in the fringe. Other projects are perhaps more passive, and simply create and promote an environment that provides mental and physical stimulation. The Pennyhooks Centre incorporates wildlife habitat improvements that are intended for its autistic visitors, though more general projects may reap comparable rewards for a wider cross-section of society (Penn Associates, 2002: GW5).

In this section, the word 'recreation' has been used to denote a range of potential experiences of the fringe: for health, education, skill development and general leisure. For many, these are the possible sociocultural functions of the fringe. But as we have suggested, they are just one dimension. These are the functions that government would like to see promoted and further developed. But there are other ways in which the fringe is experienced and 'consumed': for example, by those looking for musical or artistic experiences, motor-sports, sailing opportunities (which are also highly educational), or just an afternoon at the cinema. But none of these experiences are possible without access to the fringe. A network of trunk roads will connect nearby towns to commercial leisure and retail complexes, but the same network – truncating the fringe – may impede access to the open spaces within this fragmented landscape.

Accessing recreational experiences

Indeed, if the fringe is a landscape of experience, populated on a temporary basis by a community of users, then a critical issue at the fringe is one of access. Access has two dimensions: first, how do those who do not live in neighbourhoods abutting the fringe physically access these areas, on foot, by bike or via public transport? And secondly, once there, how is access to the land itself – much of which is in private ownership – actually secured? The second issue is given briefer treatment than the first. On the issue of 'securing access', Shoard (1999) offers an exhaustive analysis of the impediments to public access in her *Right to Roam*. On the specific issue of accessing the fringe, some of the potential problems of gaining legal access to open land at the fringe are suggested within planning policy itself. Planning Policy Guidance Note 17 (*Planning for Open Space, Sport and Recreation* [ODPM, 2002b]) is the most recent planning policy statement mentioning access and recognizes the special potential of the 'urban fringe' for sport and recreation provision arising from its proximity to urban areas. However, as far as provision for informal outdoor recreation such as walking and picnicking is concerned, the PPG merely encourages local authorities to develop areas of managed countryside, such as country parks and community forests. It does not urge authorities to improve public access to undeveloped land, including farmland and woodland, in the fringe in ways that would enable access to coexist with existing land use activities. In this respect, the Guidance Note's failure to refer to the improvement of rights of way in the fringe might be seen as a step back from the government's previous advice in this area. PPG 17: *Sport and*

Recreation issued in 1991, talked of the desirability of increasing public access to open land in the fringe and of improving the rights of way network there. It is possible that the government will reflect the goals of its 'liveability agenda' in future advice to local planning authorities and that a new PPS 7 will call for stronger action with regard to legal access of private land in the fringe. Indeed, further advice might well be offered on the local development of 'rights of way improvement plans', which may enable authorities to improve public access within the fringe.

Under the Countryside and Rights of Way Act 2000, every local highway authority has to prepare and publish a rights of way improvement plan within five years and review it at intervals of not more than ten years. Local authorities will be expected to consider the extent to which current rights of way meet expectations and demand. And these expectations may well reflect the government's liveability agenda: strategies aimed at environmental regeneration (such as the Communities Forests Programme, or the more recent 'greening strategy' within the Thames Gateway) would be rendered meaningless if people were to be prevented from accessing these greened areas. A number of mechanisms are currently available for securing access to open space, and these have been updated in the last five years (with the development of more access plans and agreements). Country Parks, for example, are one of the best known tools for providing recreation in the countryside, including the rural-urban fringe. The tool was devised in the 1960s as a means of providing 'honey pots' close to towns that could absorb large numbers of visitors with their cars and thus protect what were considered more vulnerable rural areas beyond (Shoard, 1997: 319–325). In the fringe, some of these parks have been established on the sites of old mineral workings (see Chapter 5). Many country parks have been created, and 65 per cent of all the parks are situated in the rural-urban fringe (Collins, 2003: 252). Two recent reports (one by the House of Commons Select Committee on the Environment, Transport and Regional Affairs in 1999, and the other by Sport England in association with the Countryside Agency and English Heritage in 2003) have drawn attention to the generally declining state of some of the parks and, in particular, of buildings within them. It is argued that more money is needed in order to maintain car parking and to provide basic facilities.

Country Parks provide open access to designated areas, but these may not be conveniently located for all communities living close to the fringe. For many, more general access is desirable, allowing them to cross a fence onto nearby open space and simply walk the dog or play football with their children. Access improvement plans may well become the principal tool for creating such opportunities in the future, but in the recent past, Section 106 Agreements have been used with some success. These agreements – based on Section 106 of the Town and Country Planning Act 1990 – allow local authorities to attach obligations to new planning permissions, perhaps in order to secure community facilities. Shoard (1999) cites two instances in which Section 106 Agreements have been deployed to secure new public footpaths when planning permission was granted in the mid-1990s for golf courses on the edge of Theale in Berkshire, and north of Newbury (Shoard, 1999: 34). No doubt many other examples could be found. When the London borough of Hillingdon granted planning consent for a business park near Heathrow Airport in the early 1990s, it secured through planning gain a pay-as-you-play golf course, football pitches, children's play areas and a network of footpaths. Section 106 Agreements linked with planning consents can be used in a variety of situations to deliver a variety of outcomes. However, they are restricted to situations in which planning applications are submitted for developments likely to generate a fairly large amount of income and in which the developer is prepared to agree to the provision of the particular facilities that are requested.

Where legal access to open space has not been secured, illegal access may prevail, sometimes with landowners turning a blind eye to people simply picnicking on rough grazing land adjacent to a housing estate. In other instances, landowners may attempt to derive an income from organized events, renting land for festivals or for activities such as paintball or quad-biking. The operators of airfields may close part of the field to flying and rent land to the organizers of a Sunday market or motor sports. There are a variety of ways in which the fringe can be accessed and experienced.

But the bigger problem in some fringes is one of physical access: major motorways may present a physical barrier to access, or the 'best' open land, most suited to informal recreation, may be some distance from the urban edge. Public transport provision can be particularly limited within the fringe. There are of course exceptions: residents of northeast London enjoy good public transport services through the Lee Valley Regional Park. But in general terms, where public transport routes exist, they tend to pass through the fringe rather then serving it. Osment (1998) argues that a lack of access to the fringe will limit the extent to which visions for a 'multifunctional' fringe can be realized (see Chapter 1). Without access, some fringe areas will remain characterized by closed-off fragmented spaces: people will not be able to experience these areas and thus their 'sociocultural functionality' will be limited. Some fringes do have networks of lanes, footpaths and bridleways, but others have transport networks designed wholly with private cars in mind: networks that link together business parks, out-of-town shopping and so on. Often, peripheral movement routes such as ring roads have been built that cut off traditional radial patterns of movement from the urban to the rural; and areas between peripheral routes and the urban edge have been in-filled with development mainly accessed by car. This problem has resulted, in some instances, in the development of policies aimed at re-integrating town and country. The desirability of having access to 'countryside' was demonstrated in 2005 in a survey undertaken for the Halifax Building Society which concluded that 'easy access to the countryside is the most important factor in deciding where to live for more than a quarter of UK residents' (Planning, 2005: 4).

Some authorities have already reacted to this desire and instigated the development of 'Greenways' designed to give people easier access to the fringe. 'Greenway' is a term used in mainland Europe and the UK to describe communication routes that have been developed either for recreational purposes or undertaking daily trips (such as getting to work or college or going shopping) for non-motorized traffic along pre-existing route ways, in particular disused railway lines and towpaths (Figure 6.7). A report published by the European Greenways Association (2000) singles out as an example in England a Greenway established on a disused railway line running between Bristol and Bath, which became a Greenway cycle route under the coordination of 'Sustrans' (a sustainable transport charity that 'works on practical projects to encourage people to walk, cycle and use public transport in order to reduce motor traffic and its adverse effects' [www.sustrans.org.uk]). Much of this particular route runs through fringe areas. More than one and a half million journeys along it have been recorded in one year – made not just by cyclists but also by pedestrians and people in wheelchairs. Property values nearby have apparently increased since the Greenway was established (European Greenways Association, 2000: 37), perhaps confirming the residential preferences noted by Halifax Building Society. In a similar vein, Bickmore Associates (2003) have focused on the issues of access and enhancement in the fringe, arguing that in relation to undeveloped land in the inner fringe, there are opportunities to develop green space strategies – that incorporate habitat enhancement – which extend from the heart of built-up areas out into the rural-urban fringe. These could, for example, run alongside rivers or transport corridors, extending existing habitats, providing new opportunities for recreation, and improving access (Bickmore Associates, 2003: 3).

Access to the fringe is clearly an important issue and practical steps are being taken to secure this access and to provide the means by which people can venture into the fringe. Obviously, if the fringe is credited as having a sociocultural function, then those deprived of access will be considered socially excluded: they will be unable to enjoy the social and recreational

6.7 Cycling in the fringe, Birmingham (Countryside Agency, Julia Bayne).

opportunities that exist in the fringe. They will also be denied access to the health benefits of walking and cycling, especially if routes into the fringe are unsafe or services irregular; and they could potentially be barred from jobs located in the fringe if public transport is poor and accessing such jobs is entirely dependent on car ownership. Thus improvement in public transport – or the creation of Greenways – is critical not only in order to make sites accessible for recreation, but also to allow people to take up employment in the new economy of these areas (Chapter 5). Recent work by Land Use Consultants (2004) has looked, in particular, at 'the policies, issues and opportunities related to open space and countryside around London, focusing particularly on the role of agriculture and public accessibility issues'. The study area embraced all of London's green belt and extended out as far as the M25 orbital motorway (Land Use Consultants, 2004: 1). On the issue of public transport accessibility to open green space in London's green belt, the report gave a mixed picture:

> Radial routes are very important for accessing the Green Belt and there are a number of areas where rail links serve accessible open space well with stations near to a block of Green Belt providing a gateway to that area . . . For example, many of the sites within the Green Belt in the south of Croydon are accessible by rail, and sites in the eastern part of Croydon are well served by a rapid transit line. However, some sites are poorly served by rail, for example many sites in Bromley and Bexley are not well served. Epping Forest and the Lea Valley, the two largest areas of POS [Public Open Space] within the study area, are both well served by several railway stations.
> (Land Use Consultants, 2004: 41)

Established recreational areas in the fringe tend to be more accessible; it is those areas of less established potential that are more disconnected and may not be accessible to groups experiencing 'transport poverty', including the young, the elderly and the unemployed. As more people access the fringe by car – especially more affluent homebuyers seeking proximity to open spaces – demand for public transport may decline, making already less profitable fringe bus routes increasingly difficult to sustain. Inaccessibility is a major barrier to realizing the full 'sociocultural functionality' of the rural-urban fringe. But, on the other hand, many experiences at the fringe are tied exclusively to the car: these include the leisure complexes and the other more formal – and often less 'green' – recreational activities described at the beginning of this chapter. The fringe may well be a potential environmental resource – a 'classroom', a 'health centre', a 'nature reserve' and a 'place for sustainable living' (Countryside Agency, 2005) – but it is also a space for other types of experience: for leisure activities that require large amounts of indoor or outdoor space, and that are frequently accessed by car. This is an important dimension of the sociocultural fringe, but also an aspect of the service function that has grown within the economic fringe. There are clearly problems of compatibility between service sector leisure and informal recreation: the former has been built on the densification of the peri-urban road network during the last half century, and is a major consumer of open green space (for golf courses, or for parking). Spaces for the latter, however, may become less accessible (and less attractive) if the landscape becomes increasingly truncated and fragmented. This problem illustrates one important barrier to achieving integration between functions.

Landscape can be 'made' (see Chapter 2) to favour particular uses: the fringe has been made by the car – to a large extent – and now supports a largely car-based economy. For other functions to secure a place in the fringe – or to realize full potential – they need to find ways of integrating with the landscape that already exists. This might mean, for example, Greenways linking to transport hubs that themselves are connected via public transport to nearby 'gateways' at the urban edge or within towns and cities. In Stratford-upon-Avon, a Greenway along a five-mile section of the former Honeybourne railway line (closed since 1976) has three access points at Severn Meadows (a mile from the town centre), Milcote and Long Marston. Each has a car park and is accessible by bus. This is a fairly small Greenway beginning within a fringe that has not been heavily scarred by transport development. However, this type of integration between cars and the Greenways is likely to prove necessary in some instances, as is the more general need to integrate 'green' recreation with shopping, and perhaps even with the multiplex cinemas mentioned earlier. But this may not be an option contemplated by the agencies currently leading on the fringe agenda in the UK. The Countryside Agency (which will see its landscape, recreational and access functions move formally to 'Natural England' in January 2007) and the Groundwork Trust (a federation of local trusts promoting 'environmental action') have tended to concern themselves only with 'green' or 'environmental' experiences within the fringe, and hence commercial development is rarely integrated with environmental projects. This separation of 'green' and 'commercial' was challenged with the development of the City of Manchester Stadium for the 2002 Commonwealth Games. Though not in the fringe, the area of east Manchester in which the stadium is situated is physically cut off from the city centre by the mainline railway running into Piccadilly Station (to the west). The creation of a green walkway between Piccadilly and the Stadium, taking in sections of the Rochdale Canal, was designed to break down this physical barrier. Its success is difficult to judge, though supporters of Manchester City FC – which now occupies the Stadium – are encouraged to walk the 20-minute route to the fixtures. In Manchester, a consortium of interests came together to realize a very multidimensional vision. But in the fringe, the fragmented politics noted by Carruthers (2003) tends to persist. The hiving off of the Countryside Agency's 'softer' landscape functions to the new agency – with its more economic functions being devolved to the Regional Assemblies – may make it more difficult to integrate more environmental objectives with commercial ones, not that the two need to be kept separate.

Conclusions

In this chapter, we have depicted the fringe as a landscape of experience. The idea of a 'sociocultural' fringe may suggest a residential function for the fringe that builds on the goal of trying to strengthen residential and environmental linkages. More than a hundred years ago, Howard (2003 [1898]) promoted the view that town and country might be better integrated, affording people the dual benefits of town living and proximity to nature. The desire to have a mix of convenience and open space remains intact to today, with the Halifax Building Society demonstrating people's continued desire to have easy access to the countryside. But the fringe is not a focus for residential development, partly because of planning restriction, and also because those areas subject to house building on a grand scale are consequently no longer thought of as fringe. However, some residential developments have sprung up at motorway junctions and might be viewed as evidence of planning's failure to promote more sustainable lifestyles. Clearly there are a range of factors resulting in the development of what we have termed 'edge centres'. The principal suggestion in this chapter is that these developments often represent the antitheses of good planning and design, but that a different development model – that of 'cluster housing' – might provide an alternative model for hub development that avoids the anonymity of further suburban development.

However, the argument that the fringe, by definition, is non-residential led us to conclude that the 'community' of the fringe tends, instead, to be a community of users who experience the opportunities in a variety of ways. Clearly a division exists between those experiences tied to the new economy of the rural-urban fringe, and those seeking to exploit the more intrinsic qualities of the landscape. The development of the fringe over the past 50 years appears to favour formal leisure accessible by car; there is a general underdevelopment of 'green' recreation and this is something that environmental groups are currently seeking to remedy. But there are clear tensions between commercial leisure and environmental recreation: commercial leisure seems to have the upper hand as it is more accessible to its particular user group, concentrating at transport hubs accessible by car and bus and, in the case of parkways, by train. Opportunities for environmental – or just less formal – recreation, on the other hand, are more dispersed, making access more difficult. Mechanisms have existed for some time to open up key areas – especially through the designation of country parks – though these tend to concentrate demand and visitor pressure. One way forward would be to integrate formal and informal recreation, creating common access nodes. But this is not easy to achieve given the different interests that preside over – or seek to influence – these functions. Aspects of the 'sociocultural fringe' are difficult to integrate and there are clearly competing demands for very different kinds of recreation in England's fringes.

7

An ecological fringe

Introduction

We have already alluded to a certain ecological richness in the fringe: the unimproved grasslands where people walk their dogs (see also Chapter 9), the woodlands which have become battlegrounds for paintball, the old airfields with many hectares of open space, and the derelict buildings of the Industrial Revolution (see Chapters 2 and 5) all have an unplanned wildness that suggests potential diversity in flora and fauna. Shoard (2000) has called the fringe a 'refuge for wildlife driven out of an increasingly inhospitable countryside'. In *Edgelands* (Shoard, 2002) she suggests that the diversity of plant life often found in fringes is commonly a product of local conditions – the mix of soil types, the 'deposition of builders' rubble', and horse grazing (ibid.: 125) – which often come together in a way that is 'highly unlikely in nature'. At Molesey Heath in south-west London, Shoard identified a total of 311 species of flowering plants and ferns (ibid.: 125) but she adds that the Heath

> is not unusual in the rich diversity of its plant and animal communities. Other interfacial wildlife hotspots include the land around Beddington Sewage Works near Croydon, which is one of the top bird watching sites in southeast England, while all five species of Grebe occurring in Britain from the familiar great-crested to the rarely seen red-necked have been recorded at Stonar Lake on the edge of the small town of Sandwich in Kent.
>
> (ibid.: 126)

The latter is an 'extensive, neglected water' resulting from quarrying at the site more than a hundred years ago. Other similar oases are identified by Shoard across Britain and it is explained that 'stretches of neglected rough space in towns can be rich in wildlife, particularly if . . . deposited waste has contributed to the underlying rock and soil' (ibid.: 128) giving it unusual textural, nutrient and pH characteristics. The fringe regularly provides an abundant source of such 'rough space' because land is fragmented, neglected and rarely farmed intensively. It may also have proximity to gardens, another 'increasingly important wildlife refuge' (ibid.: 129). Shoard – perhaps the key commentator on the 'ecological fringe' – summarizes and explains the ecology of the fringe in the following terms:

> Interfacial sites often enjoy biological diversity partly because they are ignored. Being ignored, they go unmanaged. The clutter of the interface, which would be tidied out of sight by those concerned with creating an acceptable landscape there, often enhances wildlife by creating new niches that wild creatures can exploit. Throw an empty milk crate into a lake and while it may look untidy, fish will swim in and out of it and use it as part of their ecological world. Black redstarts nest in the brickwork of derelict buildings. So while town parks are grassed over for ball games and our national parks overgrazed by sheep, these truer wildernesses are allowed to find their own accommodation with nature, evolving silently and unhindered.
> (Shoard, 2002: 129)

Shoard has championed the view that it is the wildness and scruffy nature of the fringe that should be celebrated, challenging the landscape conventions described in Chapter 4. The rural-urban fringe has a mosaic of open and derelict spaces often truncated by roads and punctuated by transport hubs (see Chapters 2, 5 and 6). But it is not only the derelict and neglected spaces that provide refuges for wildlife: even some of the gardens around business parks, land straddling railways and motorways, and fringe cemeteries offer a sanctuary for wildlife from more 'conventional' countryside that endures the sterilizing effect of intensive farming. It is the value of this landscape that we are most concerned with in this chapter, the need to afford it some form of protection, but perhaps most importantly, the case for avoiding excessive planning intervention. The ecological fringe described by Shoard is, for some commentators, the true essence of the fringe and 'one could argue that it is a contradiction to try to intrude the dead hand of the planner into something whose character is to be free' (Shoard, 2002: 140).

And yet, Shoard suggests, planning – of a type – is needed at the fringe. This should not be planning that aims to 'sanitize' these landscapes, but planning that aims to provide access and to facilitate appreciation: 'in the context of the edgelands, we need to see the planner not as a shaper of an entire environment but as a handmaiden, who helps along a universe he or she does not seek to control' (ibid.: 140). Hence, Shoard favours the creation of networks of green spaces, with programmes overseen by small teams of dedicated people whose task it would be to talk with landowners, investing their time in negotiating and improving access (ibid.: 141). The general strategy would be to go with the grain of fringe development: to allow organic change of the type described in Chapter 2. But Shoard's 'handmaiden' will have to be more than a passive guardian of the landscape. The fringe is an arena of sometimes intense land-use conflict and there is rarely agreement on the balance that should be struck between development needs and conservation (Osment, 1998; Bickmore Associates, 2003). Neglected rough space in towns – celebrated by Shoard (2000, 2002) and by Kendle and Forbes (1997) – has become a key focus for housing land release as government pursues its 60 per cent target for future house building on 'previously developed land' (PDL: see DETR, 2000a). The fringe faces a similar pressure, not only from housing development (though this does occur, especially around transport hubs: see Chapter 6), but from a range of new economic uses that frequently consume large tracts of open space (see Chapter 5). In the remainder of this chapter, we consider the balance between development and ecology in the fringe, ending with further consideration of the role of planning at the edge.

Developmental versus ecological agendas

In a recent examination of the housing planning question and its political context, Murdoch and Abram (2002) have suggested that the planning for housing debate in England (how many homes should be built, of what type and where) is shaped by the tension between developmental (pro-house building and general housing provision) and environmental interests (anti-development in many instances). The same debate – cast as development versus environment, or in this instance, development versus ecology – is evident in all areas of land-use policy. Sometimes, the environmental agenda may become a convenient cloak for NIMBY interests, but in the rural-urban fringe, where land-use change is frequently unchecked by communities, the battle-lines are more clearly drawn between those who view the fringe as possessing only potential development value, and those like Shoard, who recognize intrinsic and unique ecological qualities. These two agendas have been at loggerheads for a number of years. It has been argued (see Murdoch and Abram, 2002) that planning in the mid-1980s was located in a broadly 'developmental' paradigm, which was gradually diluted during the 1990s by an environmental agenda. Indeed, Shoard (2002: 136) has suggested that 'large-scale development in many of our edgelands took off during the Thatcher years when planning deregulation was fashionable, with central government predisposed towards free market thinking'. Later in the 1990s, the priority of policy was to balance sometimes conflicting environmental and developmental agendas while introducing a new emphasis on the efficient use of resources, on place-making, on social balance and inclusion, and on lightening the ecological footprint of new development. These issues have been sown together into an official view of 'sustainable development', most recently articulated in Planning Policy Statement (PSS) 1 (ODPM, 2005). But in the fringe, 'the neglect and disdain with which the authorities regard the interface' (Shoard, 2002: 141) has meant that the developmental agenda has been less diluted than elsewhere.

We have already touched on the nature of this agenda at various points in this book. Conservation of the historic heritage (Chapter 3) of the fringe often results in buildings being re-used, sometimes taking on a new economic function. The post-war period saw an explosion of road building within the fringe, which has further fuelled development (and squeezed out agriculture), especially around transport hubs. This development tends to be steered towards the service economy: the retail and business parks of Chapter 5. Housing is also a part of this agenda, but to a more limited extent. That said, projections of housing growth have raised the spectre of large-scale residential encroachment into the fringe, which has the potential to wipe out existing habitats. Talk of urban extension has quietened since Labour's first election victory of 1997. The focus today is on infill development and the development of the Growth Areas set out in the 2003 Communities Plan. In the south east of England, however, many of these centres will be focused on areas of fringe: in the Western Corridor stretching from London towards Reading; around Milton Keynes in Buckinghamshire; in the vicinity of Gatwick Airport; in the Kent Thames Gateway; and very specifically, in the 'London Fringe' (South East England Regional Assembly, 2005). These areas have been earmarked for development because they include significant areas of previously developed land in fringe locations, and their allocation for housing will ensure that the region — in this case south east England — can achieve the government's recycling target. In the aftermath of the Barker Review (looking at housing need and supply in England: Barker, 2004), pressure is on to 'deliver on the Barker agenda' and to push for further housing land allocations, especially in England's southern regions. The housing debate is often couched in terms of hard choices: for many, the environmental agenda has won too much ground, and although butterflies need homes too, the priority should be on housing the human population, on controlling house price inflation and on meeting the government's objective of a 'decent home' for all UK households. An emphasis has been placed on the quantity and cost of homes, with government calling on developers to provide £60,000 homes in southern England to beat the affordability crisis, and with the Deputy Prime Minister insisting on extra allocations in Regional Planning Guidance (again in the south of England) which will be rolled forward into revised spatial strategies and will form the basis for delivering the Communities Plan. And all this development will have to be associated with new infrastructure: with new roads, schools, community facilities, shops, sewage and water treatment works, power generation and employment sites. Given limitations on infill — which itself will obliterate areas of rough open space in towns — the fringe, and the green belt, may feel the brunt of this development pressure. Some fringe will inevitably be lost in its entirety; there may be pressure to clean up other areas, especially if they are overlooked by new housing. But can government and local authorities engage in a damage limitation exercise? Much will depend on whether Shoard's view of the landscape of the rural-urban fringe — to be celebrated rather than sanitized — can win favour, and whether the existing framework for protecting wildlife in the fringe is fit for the purpose (Figure 7.1).

Box and Shirley (1999) have added their voices to a small but growing chorus of concern over the potential loss of urban (and fringe) brown fields, and the impact this is likely to have on urban ecology. Indeed, the planning system has not tended to concern itself with ecology or biology despite some

7.1 A landscape of 'unnatural' beauty, Chinnor, Buckinghamshire (Countryside Agency, R. Pilgrim).

fairly rigid controls over development in open countryside and on agricultural land. Green belt policies (see Chapter 8) and the presumption against development in open countryside (now set out in PPS 7), imply environmental protection but these tools are not explicitly 'environmental'. They seek to prevent 'urban' development in 'rural' settings, but have no concern with the ecology of the land. Indeed, the planning system, as initiated in 1947, is primarily concerned with what happens on, under or over land and not with detailed characteristics. This 'land use' emphasis has somewhat mellowed over the past 60 years, but the simple presumption remains that open fields should be protected from intrusive development (which threatens openness) and that urban sites should almost always be favoured for redevelopment. This message came out strongly in the Urban Task Force's final report in 1999, which ignored the potential of brown-field sites to accommodate biodiversity and nature conservation, focusing instead on what percentage of future development might be steered away from green-field sites. The point has almost been reached where it is no longer worth debating the ecological richness of brown-field, and fringe, habitats versus the sterility of much of the wider countryside. It is generally acknowledged that farming has wrought environmental havoc in the wider British countryside. The patchwork of small fields (the home of mixed farming) separated by hedgerows has given way to huge fields of cereals fenced off with barbed wire. Between 1984 and 1990, 100,000 km of hedgerow was lost in England. In the post-war period, chalk downland was ploughed up for crops, wetlands were drained and meadows destroyed by mechanized haymaking. A loss of habitat resulted in a loss of biodiversity and in the creation of a monotonous landscape, as anonymous as the rural-urban fringe if it were not for the villages and medieval market towns that still punctuate the countryside. But planning is not there to question the wisdom of sustaining practices that are inherently harmful. This is all decided at a European level, from which farm subsidies – resulting from the Common Agricultural Policy (CAP) – rain down to ensure that Britain (but more particularly France and Germany) keep farming irrespective of the economic, social or environmental logic of production subsidy.

Part of the problem is CAP, and the need for its reform, but another critical problem for the rural-urban fringe is the view that these anarchic landscapes are simply not worth protecting. This has nothing to do with the ecology, but with conventional landscape tastes and aesthetics, which we considered at some length in Chapter 4. The planning system serves a majority view and this view is that green fields and 'open countryside' are more quintessentially English than the functional landscape of the fringe. It is of no concern that there may be greater diversity in animal life within a single square metre of rubbish heap than in the same area of ploughed field. Shoard suggests that it all boils down to a question of understanding: 'instead of seeing the interface as a kind of hellish landscape to be shunned, we should celebrate it', adding that 'if people were encouraged to understand this world more, they might feel less alienated and puzzled by the circumstances of their lives' (Shoard, 2000: 142). Shoard's impassioned writing on the 'edgelands' is both a plea to understand the wider fringe, and to break with landscape dogma. Biodiversity is not something found only in national parks or designated Sites of Special Scientific Interest (SSSIs); rarely seen birds nest in abandoned buildings, not just in isolated inaccessible cliffs; and 'green' does not necessarily equate with biodiversity. However, although policy might be reinforcing a stereotype of the 'perfect' landscape, there is a case for retaining openness – and hence favouring fringe development – that is quite separate from the issue of biodiversity. Perhaps the most sensible way forward is to conclude that ecology is one factor that should be considered when allocating sites for development in Local Development Frameworks (replacing Local and Unitary Development Plans after the 2004 Planning and Compulsory Purchase Act). As it stands, local authorities have been called upon to adopt a 'sequential approach' to housing allocations – ensuring that brown-field options are exhausted before considering the release of green fields – but should this approach be challenged? Box and Shirley (1999) confirm that biodiversity is ignored in land allocation mechanisms, but suggest that it needs to be factored into the planning process: every site proposed for development or redevelopment should be treated on its own inherent merits rather than merely classified as brown field or green field, with subsequent decisions made on that limited basis. An assessment should be made of its value prior to a decision being reached on its future (ibid.: 309).

But the harsh reality at the fringe is that many policy makers are unlikely to rein back development proposals for purely ecological reasons: there has to be some social or economic pay-off. In Chapter 6 we examined how the fringe might be conceived as a landscape of experience and noted that it

144 *An ecological fringe*

7.2 Wetlands, Stedham Common, West Sussex (Countryside Agency, P. Greenhalf).

provides opportunities for learning: Shoard adds that 'its dereliction stimulates the imagination' (Shoard, 2002: 130). Children play in the fringe and in doing so, also learn. This gives it a social value quite separate from its redevelopment potential; another social value relates to health. The fringe can be an ideal place to walk or play informal sports – the Sunday kick-around – assuming it is accessible. But the general idea that open space and the odd patch of dereliction has some social value is different from accepting that it might be 'better' to leave an area of wetland undrained for the sake of maintaining biodiversity, rather than to develop the site for an Ikea superstore, and take advantage of excellent transport links (Figure 7.2). But the decision, of course, is not which is better but how can both be accommodated (within the fringe, if not on that particular site)? And is there a suitable framework in place that allows the planning system to recognize and manage landscape externalities?

Accommodating the ecological fringe

Scientists – and especially biologists – have been the first to recognize the richness of wildlife often present in the fringe, and also the broader role that fringes play in supporting urban ecosystems (Harrison et al., 1996; Douglas and Box, 2000; Ling, 2001). However, official policy statements or edicts rarely mention the rural-urban fringe, let alone its potential ecological function. This is true in Planning Policy Guidance Note 9 (*Nature Conservation*: DTLR, 2002) and in the Department for the Environment, Food and Rural Affairs' comprehensive countryside survey *Accounting for Nature* published in 2000 (DEFRA, 2000). It may be the case that policy needs to catch up with current thinking on biodiversity: it was not until the 1980s that the rich ecology of fringe areas became widely recognized (Gilbert, 1989; Shoard, 2002; Whitehand and Morton, 2003). Shoard (2002) continuously reminds her readers that the fringe has a 'hidden richness', which has apparently remained hidden from policy makers for longer than it has for many nearby communities. Although local anecdote may reveal the extent of ecological richness in pockets of fringe, there are no hard data of the kind the government frequently requires before it will move to action. However, this situation is changing. An indication of the possible ecological value of some interfacial land is contained in the study *The Changing Flora of the UK* (Preston et al., 2002) in which wildlife 'hotspots' (ten kilometre squares with the highest numbers of plant and animal life) are identified within each country of the UK. While popular perception might place such squares in, say, the Scottish Highlands or the Snowdonia National Park, none of the hotspots were found in such remote areas. In Wales it was the kilometre square west of Llanelli, which includes post-industrial sites such as disused steelworks and abandoned fuel-ash lagoons, which are of highest botanical interest – as well as sand dunes, ancient woodlands, acid heathland and other habitats. Scotland's hotspot was at Dunbar in East Lothian, in the Edinburgh sub-region; England's was in Dorset, including the town of Wareham together with heathland, chalk grassland and the lower reaches of rivers; Northern Ireland's was a square that includes most of Belfast: residential and industrial areas including fragmentary coastal habitats. Clearly, such data challenges convention and presents policy makers with two questions: first, is it desirable to retain and protect these habitats? And second, how might this be achieved?

There is of course a general framework for protecting wildlife that can be applied to habitats in the fringe. This tends to pick out what are considered to be the best examples of certain kinds of habitat – for example, wetlands, chalk downland, meadows, or ancient woodland – and then affords them statutory protection. Sites of Special Scientific Interest (mentioned above) perform exactly this function and cover roughly 6 per cent of England's total land area. These are a powerful and fairly restrictive tool – which hand English Nature the casting vote on planning and development issues – but have rarely been applied in the fringe, for the reasons set out above: the special scientific attributes of the fringe are rarely acknowledged. This may be because land looks almost to be abandoned and may not attract the attention of local authorities. In many instances, it is more likely to be viewed simply as degraded land with future development potential rather than open land (Kendle and Forbes, 1997) of ecological interest. A general ignorance pervades thinking on the fringe:

> We know surprisingly little about the open spaces that we live among. Even local governments, charged with responsibility for management of much of the land, have until very recently been able to get by with an amazingly vague picture of the land they look after. Certain changes have occurred in the UK that will lead to better information supply, most notably the development of land registers, local authority environmental strategies and the requirement for Compulsory Competitive Tendering [in the recent past] in local authorities departments. . . . However even this data-gathering process will leave us with a very incomplete picture.
> (Kendle and Forbes, 1997: 1)

Within the general framework, there are also management and 'environmental regeneration' tools that can be applied locally. These include, amongst others: biodiversity action plans, local natural reserves, community forests and Countryside Stewardship schemes. Some of the key features of these tools are set out in Table 7.1. A critical point that emerges is that while some mechanisms simply attempt to protect ecological qualities, others try to integrate them with other landscape functions. This is a crucial issue at the urban edge where, as we have seen throughout Part Two of this book, 'multiple fringes' vie for position. Any tool designed to reflect the interests of the ecological fringe must be able to do so in a context of intense competition for land. Biodiversity Action Plans (BAP) have no statutory underpinning and vary in effectiveness from place to place: they are very much dependant on cooperation between the partners taking forward these plans. Local Nature Reserves can be designated in the fringe, but only if the local authority has a proprietary interest in the land: clearly a major limiting factor. Countryside Stewardship is tied to CAP and is, arguably, the product of a fairly limited and limiting view of how land should be used. But Community Forests, in contrast, seem to offer a more useful model of cooperation in the fringe, which may allow ecological concerns to be accommodated in a framework that also reflects economic and sociocultural demands on space. Perhaps the partnership arrangements and the functional integration – that is at least sought – by Forest Partnerships might be held up as something to replicate more widely. However, the Forests themselves have fallen short of expectations on several fronts. They have been more successful in instigating environmental regeneration in areas where at least some land is in public ownership: this experience emphasizes the need for public and private interests to work together for a variety of reasons, and not least to integrate economic with environmental or ecological goals. The 'working principles', including partnership working, which might guide future planning and management at the fringe, are returned to in Chapter 9.

Table 7.1 Landscape protection and 'environmental regeneration' tools

Biodiversity Action Plans

The Rio Earth Summit of 1992 led to the 'Biodiversity Convention' and global targets for biodiversity. In response several local authorities in England have produced local 'Biodiversity Action Plans'. The Department for the Environment, Food and Rural Affairs has tried to regularize the process by establishing a framework for biodiversity in the document *Working with the Grain of Nature* (DEFRA, 2002). However, the usefulness of local plans varies a great deal from place to place. In a report published in 2000 looking at 'UK Biodiversity', the House of Commons Environment, Transport and Regional Affairs Committee remarked that, 'it was a source of disappointment to many witnesses that the biodiversity action plans had little statutory underpinning' (House of Commons Environment, Transport and Regional Affairs Committee, 2000: Para. 26). Clearly local authorities' powers to further biodiversity concerns are strictly limited: much depends on persuading other organizations to take action and on reacting to what others are doing, for example seeking planning gain involving action on the biodiversity front when applications for planning permission are made. It seems clear that plans would be more effective if they were drawn up by the owners of land. The Highways Agency, for instance, has drawn up a general biodiversity action plan which sets targets to help plants and animals on its own land over the next 10 years and commits the agency to investing £15 million to developing biodiversity nationally (Highways Agency, 2003). This strategy is likely to have an impact on the fringe where the highways network is particularly dense.

Community Forests

The Community Forest Programme was initiated in 1989 with the purpose of creating well-wooded landscapes in and around major urban areas. The programme was designed to bring together local communities and other key stakeholders with a view to achieving a wider array of economic and regeneration goals. Twelve Community Forests were designated in England in two stages: the first designations were made in 1989; and a second set in 1991. A number of concerns, however, have been voiced in relation to the implementation of the programme. These concerns relate to the problem of steering 'environment regeneration' on land that is not in public ownership, and an apparent lack of leadership. Despite the fact that over 80 per cent of land in Community Forests is privately owned, research from 1995 has shown that the majority of planting has taken place on the remaining 20 per cent of public land (Counsell, 1995; Shoard, 2002). This means that while the public sector could play a much stronger role, the overall pattern of land ownership in England makes private sector involvement the key to the long-term success of the programme. Furthermore, the Community Forests have been set up as a form of partnership comprising the Countryside Agency, the Forestry Commission and local authorities in the Community Forests areas. Typically, funding is divided equally between the Agency and local authorities, but in terms of their effective implementation Community Forests are dependent on local authorities to drive forward and give more general support to their development.

Table 7.1 continued

Local Nature Reserves
Another policy tool available to local authorities is the 'local nature reserve'. Large numbers of local nature reserves have been designated in recent years and some are located in the rural-urban fringe: these include the extremely successful Lewes railway lands in East Sussex. However the requirement that the local authority hold a proprietary interest in the land in question before a local nature reserve can be designated is probably one of the main reasons limiting more extensive use. Local nature reserves offer the additional benefit that the local authority can not only protect the site but also carry out proactive management including the management of public use. Where land is in public ownership, ensuring access and therefore a degree of 'multifunctional' use is a fairly straightforward task.

Countryside Stewardship
'Countryside Stewardship' is the main scheme offering voluntary payments to landowners and managers who agree to engage in 'not-for-profit' activities such as wildlife and landscape conservtion and public access. The Department for the Environment, Food and Rural Affairs' *The Countryside Stewardship Scheme: Traditional Farming in the Modern Environment*, the principal publicity document regarding Stewardship, states that 'anyone who owns or manages land can apply for stewardship, including voluntary bodies and local authorities' (DEFRA, 2001: 4). Despite this breadth of appeal, the entire document is geared towards the farming community. But it is the thrust of the documents rather than the precise terms of reference of the scheme that is likely to exclude much fringe land that could otherwise benefit. The document does not single out for special mention any of the types of open land in the fringe that might contain valuable wildlife sites that are not farmland, such as golf courses, horse paddocks, rubbish tips, sewage works and unmanaged land in general. The owners of these types of land will probably be unaware of countryside stewardship unless they happen also to own farmland. In fact, one of the categories of site in which Countryside Stewardship aims to foster conservation is 'countryside around towns', but again, the description focuses on farmland and woodland.

Often, projects that bring together ecological and other concerns do so in an ad hoc way: local solutions are simply applied to local environmental and economic issues, though some of the tools listed in Table 7.1 might prove useful. It is also the case, as we noted earlier, that improvements to biodiversity in the fringe are rarely made with that objective in mind. More frequently improvements occur on the backs of other activities. One particular category of such improvements is currently occurring when the Environment Agency carries out flood defence and flood alleviation schemes to protect urban areas. These schemes are necessarily close to urban areas as they are intended to protect them. But the Environment Agency has a duty to promote water-related recreation and its capital projects therefore would be expected to reflect this. One example – mentioned in Chapter 5 – is the Jubilee River which was completed in 2002. Here, 14 km of artificially created channel takes floodwater from the River Thames in a brand new channel and deposits it further downstream. The Agency has not constructed a concrete drain but has tried to re-create wildlife habitats that have been lost along the Thames. In addition, it has created wetlands and scrapes at one particular site. It has also established a permissive path almost along the entire length of one side of the channel, accessible to wheelchair users as well as people on foot, and this is to be converted to a public right of way. About a quarter of a million native trees and shrubs have been planted and a 25-year management plan drawn up to improve new and existing woodland. At Jubilee River these improvements have taken place on the back of major capital works costing £100 million.

Another – albeit far smaller – development in the fringe that has tried to marry a developmental with an ecological agenda can be found at Van Diemen's Land near Bedford (in the Forest of Marston Vale [FMV]), where a partnership comprising the Forestry Commission, DEFRA and the Marston Vale Trust has tried to create a community woodland and new habitats as a context for new retail and road improvements. Here an 8 ha area of former arable land has been developed as a large-scale distribution centre for major retailers including Asda and Argos. The distribution centre is linked to road improvements in the area, including the construction of the Bedford Western Bypass and improvements to the A421. Part of the site functions as a 'floodwater storage lagoon' and it was therefore necessary for the partnership to enter into talks with the local Internal Drainage Board – responsible for flood management through watercourse maintenance – to ensure that woodland planting would be compatible with the operational function (i.e. drainage) of the land. This is not an example of an existing habitat being protected or enhanced, but an entirely new habitat being created. It illustrates that ecological and economic functions can coexist, though the driving force behind the scheme was the retail distribution development which, through a planning gain agreement, provided the lion's share of funding for the scheme. In this case, it was possible to combine tree planting with the floodwater storage function to establish over 6 ha of 'wet woodland' (a local Biodiversity Action Plan 'target habitat') with full public access. An additional benefit – listed in an assessment of the scheme undertaken by an environmental consultant in 2002 – is that the tree planting provides 'a shady route for the Milton Keynes to Bedford cycle route that will run adjacent to it'. Local residents have also become involved, buying saplings and planting them on the site (Penn Associates, 2002: Scheme FMV1, no page numbers). According to the consultants' report, the scheme has contributed positively to forestry, biodiversity, landscape change, integrated planning, sport and recreation, 'active citizenship' and 'access for all'. Perhaps the most interesting

150 *An ecological fringe*

aspect of the project was the involvement of the Internal Drainage Board (there are three boards in Bedford known collectively as the 'Bedford Group'). We have already seen in Chapters 2 and 5 that statutory undertakers have an important place in the landscape of the rural-urban fringe, providing and managing the 'urban dowry': i.e. the essential services that are crucial to the functioning of towns and cities, and also to their future economic growth. The Van Diemen's Land Scheme would not have been possible without the involvement of the Internal Drainage Board and we have noted already that it is far easier to achieve ecological objectives and enhancements where the owner of the land is directly involved. The commitment of the Highways Agency to its national Biodiversity Action Plan may have important implications for roadside habitats throughout the rural-urban fringe. Clearly, statutory undertakers – the drainage boards, the water companies, energy distributors and so forth – have a role to play in 'managing' fringe land and its habitats, alongside that of the private landowners. But the undertakers, as major landowners in the fringe (private ownership is often far more fragmented), may be keener to demonstrate their environmental credentials and may be in a position to finance local initiatives on a national scale.

For some commentators, undertakers – now the privatized utility companies – have been viewed as the wreckers of the fringe, and it might seem paradoxical to attribute to them a more positive, environmentally conscious, role. In Chapter 2, for example, we noted Nairn's (1955) attack on 'subtopia' which was viewed essentially as a landscape of traffic, roundabouts and 'things in fields'. Thirty years later, Punter (1985) pointed out that many of

7.3 A landscape serving the city, Romney Marshes, Kent (Countryside Agency, Martin Jones).

these 'things in fields' – electricity sub-stations, pylons, sewage treatments works etc. – were exempt from planning control and the products of a number of 'unwitting agents of subtopia': local authority departments (highway engineers, public works departments, etc.), statutory undertakers (electricity and water companies, etc.) and government ministries (the Ministries of Defence and of Transport: for road schemes and military airfields). Today, water, railway and electricity companies often retain major assets in the fringe, with responsibilities for marshalling yards, water reservoirs and electricity generating stations or sub-stations, all of which have a major aesthetic and environmental impact on fringe landscapes. The National Grid Company, for example, operates the high voltage electricity network in England and Wales, which comprises some 7000 km of overland line, 21,000 pylons, 650 km of underground cable and more than 300 sub-stations. In addition, the National Grid manages 1,000 hectares of non-operational land (Land Use Consultants, 2002). These companies, as statutory undertakers – defined by the Town and Country Planning (General Permitted Development) Order 1995 – operate outside the jurisdiction of local planning authorities, but have demonstrated a willingness to take a proactive role in environmental projects – especially those that create new opportunities for public access – possibly because they wish to retain a positive relationship with local communities, and to demonstrate their environmental credentials. The latter factor may be particularly important in the age of privatized utilities: companies wish to project a 'green' image to their customer base and to the government-appointed regulators, who may have the power to penalize companies who engage in environmentally damaging practices.

Environmental initiatives are also important for companies signing up to government's Corporate Social Responsibility (CSR) agenda: i.e. 'how business takes account of its economic, social and environmental impacts in the way it operates – maximising the benefits and minimising the downsides' (see www.societyandbusiness.gov.uk; accessed 15 June 2005). The CSR 'model' (see Table 7.2) suggests that 'business success and where relevant shareholder value will . . . be better delivered by those companies which are contributing to sustainable development' (European Multi-Stakeholder Forum on CSR, 2004: 8). CSR is really an emergent framework for delivering sustainability, and much of what has happened in the UK with the utility companies – and indeed, with other companies carrying forward a 'greening' agenda – predates much of the rhetoric that has accompanied the arrival of CSR. In the UK, there has been a more general expectation that companies might undertake 'voluntary actions . . . over and above compliance with minimum legal requirements'. For this reason – as well as for reasons spelled out in national legislation – these sectors and companies all have comprehensive environmental policies in place (Land Use Consultants, 2002). Much of what they already do is linked to landscape and environmental impact mitigation: but sometimes their concern is also with biodiversity. For example, we have already cited the Highway Agency's 2003 'Environmental Strategic Plan' which sets out its environmental policy and its voluntary commitments, which have resulted in the Agency working with a range of local stakeholders on ecological projects – especially at roadsides – linked to its own landholdings. Furthermore, some of the UK's utility companies are currently quoted on the Dow Jones Sustainable Development Index and appear to display a genuine interest in promoting environmentally friendly practices (Land Use Consultants, 2002), perhaps having been influenced by the CSR model.

For businesses – including the UK's utilities – reducing costs, protecting resources, differentiating your business from competitors (and therefore standing out in a sometimes crowded market), and gaining legitimacy are all important. Aside from the utilities, international companies such as Honda and Shell – associated with the motor-car and with fossil fuels – have been trying hard to stand out in recent years. Honda is apparently 'accelerating its environmental activities' and remains committed to 'handing down joy from one generation to the next' (world.honda.com; accessed 15 June 2005). Shell (and the 'Shell Foundation') has sponsorship links with the Royal Geographical Society and at a meeting at the Society's headquarters in London in 2005, reiterated its commitment to various biodiversity standards. Indeed, the company clearly views itself – along with the World Wildlife Fund and others – as part of the 'biodiversity community' (www.shell.com; accessed 15 June 2005).

But back to the fringe and to the utilities: it is clear that there are a wide range of views on both CSR and the level of commitment that companies should show to habitat protection and public access. Some companies remain concerned about the level of business opportunities created through environmental commitment; others are worried about public access that could bring vandalism and fly-tipping and might gave insurance implications (indeed, such concerns are shared by all landowners thinking about widening access in the fringe, and elsewhere). Essex Water Company (now Essex and Suffolk Water and part of the Northumbria Water Group), for example, blocked an attempt in the late 1990s by Essex County Council to secure public path

Table 7.2 Corporate Social Responsibility – key themes

➤ Reducing costs through eco-efficiency;
➤ Protecting or enhancing the resources (environmental or human) on which the business depends;
➤ Anticipating, avoiding and minimizing risk and the associated costs;
➤ Anticipating costs (including insurance costs), societal and stakeholder expectations, customer demands, and future legislation;
➤ Retaining the 'license to operate';
➤ Differentiating from, and gaining an edge over, competitors;
➤ Protecting, building and enhancing reputation particularly for branded and business-to-consumer companies;
➤ Attracting and retaining skilled and motivated employees;
➤ Learning and innovating, improving quality and effectiveness;
➤ Being an attractive prospect for investors;
➤ Improving relationships with stakeholders.

Source: European Multi-Stakeholder Forum on CSR (2004)

status to an existing track down to and along the edge of one of the company's storage reservoirs (Shoard, 1999: 94–95). On the other hand, Northumbria Water Limited – which is now part of the same group of companies – has embarked on a £70 million upgrade of the Howdon Sewage Treatment Works near Newcastle-upon-Tyne in the north of England. This site was once described as a 'horrendous wasteland of contaminated land, remnants of old building foundations, burnt out cars, syringes and the other rubbish one might expect on an abandoned brownfield' (Spray, personal correspondence: quoted in Gallent et al., 2003b: 75). As we noted earlier, in this situation some landowners have a tendency to ignore the site or to look for new development opportunities. But in this instance, Northumbria Water commissioned an ecological study to gain a baseline assessment of the site's ecological potential. Initial investigation found that part of the site was already designated as a Site of Nature Conservation Interest (SNCI) and supported a number of nationally important species and habitats. In this particular instance – though this is by no means the norm – the company then adopted a very proactive lead in subsequent habitat 'enhancement' works, possibly viewing this as an opportunity not only to enhance the local environment, but also their corporate image. They entered into a Section 106 agreement with the local planning authority and through detailed consultation with English Nature, the Northumberland Wildlife Trust, the Wetlands and Wildfowl Trust and other nature conservation organizations, they were able to produce a list of nature conservation priorities for the scheme, which was particularly important because it ensured that the project could contribute to the achievement of local biodiversity targets (set out in a local biodiversity plan – which of course, the local authority could not achieve on the site in the absence of any proprietary interest). The partnership decided that the main aims for the project would be to create a large wetland site with complimentary habitats for species-rich grassland, woodland and scrub. A main lake was designed and sluices were installed to control the water levels at different times of the year. Gravel islands were established with the principal aim of creating breeding habitats for a variety of both common and less common birds. Finally because of its fringe location and its increasing value for biodiversity, the site came to provide an ideal opportunity for use as an educational resource. Site interpretation boards and a bird hide were constructed to allow local school groups, residents and visitors the opportunity to view the wildlife present on what has subsequently become known as 'Howdon Wetlands'.

This is perhaps one of the best examples of how the 'ecological fringe' can be integrated with other competing land-use demands, and with other objectives – including public access – in the rural-urban fringe. But there are, of course, other examples. Barn Elms – 'metropolitan open land' rather than 'fringe' in the strictest sense – also illustrates how a proactive utilities company can steer a development that serves both an essential service (and hence an economic role) and an ecological function. Barn Elms, a 40.5 hectare site in south-west London, is now home to the London Wetlands Centre. Before development, the site comprised four reservoirs providing drinking water to this part of London. In the early 1990s, a Thames Water Ring Main came into

154 *An ecological fringe*

operation and the reservoirs became redundant. The local authority, the London Borough of Richmond-upon-Thames, was left to make the decision as to what to do with the site. The Reservoirs Act 1975 seemed to suggest that the reservoirs could now be decommissioned and drained. The Barn Elms site had been a Site of Special Scientific Interest since 1974 because of the variety of indigenous migratory birds it attracts; but neglect during the previous quarter of a century meant that this habitat was at risk of deterioration, particularly through silting. A further complication – or advantage depending on one's viewpoint – was that even if the reservoirs were dried out, the site could not be developed because of its status as 'metropolitan open land'. Hence a question mark hung over the future of Barn Elms, and it was decided that a compromise needed to be struck between partial commercial redevelopment (on a part of the site where this would be permissible) and maintenance of the site's ecological function. It was left to the local authority to find an 'enabling development' though Thames Water – as landowner – had a key role to play in the disposal of its now non-operational asset. The solution for the site took the following form: Berkeley Homes (Thames Valley Ltd) – a volume house builder operating across London and the south-east of England – agreed to develop a 9.03 hectare portion of the site (to the north, and adjacent to the River Thames). This 'executive development' was to be used to secure funding for a wetland project. In 1990, Thames Water Ltd began talks with the local planning authority and full planning permission was granted after the signing of a comprehensive Section 106 agreement in 1991. A phasing strategy was agreed whereby a certain number of houses had to be built before each stage of the wetlands site could be implemented (Cuthbertson, 1996). Opened in May 2000, the Wetlands Centre at the Barn Elms site hosts over 140 wild bird species, more than 300 species of butterflies and moths, 18 species of dragonflies and damselflies, four species of amphibians and a multitude of other species recorded there annually. In February 2002, 30 hectares of the centre were reconfirmed as a Site of Special Scientific Interest, apparently supporting nationally important numbers of Gadwall and Shoveler duck. It also supports a programme of learning aimed at exposing local school children to London's wildlife.

For those who see a need to integrate different functions on single sites, Barn Elms might be held up as a model for future planning. It accepts the 'harsh reality' – mentioned earlier – of having to make bedfellows of economic and environmental objectives, in many instances. But there are those who do not accept this reality, and although they might applaud what has been achieved at Barn Elms, they feel that something even better was possible and without losing part of the site to executive development:

> [Barn Elms] was 'Metropolitan Open Land' (MOL), which is a planning designation which meant it should not be built on. In fact, permission was given to build posh executive housing on some 25 per cent of the site. This so-called 'enabling development' allowed development of the rest of the site as a nature reserve to be financed. This raises serious ethical and conservation issues. The whole site

was already of considerable wildlife interest. Was it right to give up 25 per cent of the land, which was supposedly protected? While the development of Barn Elms has undoubtedly proceeded much faster with the contribution from the developer, there is no doubt that the whole site could have been developed to become a superb nature reserve, either by WWT or someone else. In the longer run, have we lost out? Is giving up 25 per cent of land that is meant to be kept open consistent with the concept of 'sustainable development'?

(West London Friends of the Earth, http://www.wlfoet5.demon.co.uk/biodiv/hammersmith/ barnelms.htm; accessed 10 February 2006)

Clearly, Friends of the Earth (FoE) raise some fundamental questions: who should meet the cost of protecting sites of special ecological interest? Is it right to bring different land uses together in a marriage of economic convenience? And what is sustainable development? Had it been possible to develop social housing on the northern section of the site, then FoE may have been less concerned over the loss of almost a quarter of the site to development, but clearly, the development had to be high value in order to carry the planning gain. It appears that, at the time, there was no other way to fund the wetland project and simply letting nature take its course may have resulted, eventually, in a different kind of habitat, and one that may not have been suited to the species that currently find a home at Barn Elms. The final question is the most interesting. In Chapter 1, we suggested that 'going with the grain' of multifunctionality – and striving to achieve a variety of economic, sociocultural, aesthetic, and ecological objectives – provides a blueprint for sustainable development. But these objectives often share an uneven relationship: the achievement of one is frequently dependent on the achievement of another. In the ecological fringe, fierce competition for space means that ecological and aesthetic concerns risk being squeezed out, making it difficult to integrate functions. The land-use planning system – a focus of concern in the next chapter – has a history of separating functions, creating a situation in which 'environmental planning' has simply concentrated on the gaps between development rather than on finding new and sustainable ways to bring together different land uses. Admittedly, the Barn Elms development was not an ideal solution: land and habitat were lost to executive housing development. But this loss might be seen in the context of an evolving planning system which, over the past twenty or so years, has gradually provided the tools – including revamped planning agreements – with which to capture some of the uplift value of development with a view to achieving wider objectives. In both Barn Elms and Howdon, positive steps were taken to realize the ecological potential of rough land and the rural-urban fringe: different functions have been combined; money has been secured for ecological projects; and business interests have shown some interest in what happens on non-commercial sites. Arguably the 'ecological fringe' remains the poor relation within the fringe family, scrambling round for recognition and acceptance: but at least there is now some commitment, through the CSR model and through the mechanisms that enable companies to deliver ecological benefits, to bring it into the fold.

Conclusions

In this chapter, we have tried to tell the story of the ecological fringe, beginning with its struggle to be recognized as an important biological, educational and aesthetic asset. Its difficult relationship with the developmental agenda – which persisted in the fringe into the 1990s – and how this relationship has slowly improved have also been charted. By 1999, the Countryside Commission – in one of its last reports before being subsumed into the Countryside Agency – anticipated that the fringe could become 'an area valued in its own right, where town meets country and where regional diversity and local distinctiveness are recognised' (Countryside Commission, 1999: 3). This has almost happened. There are instances where local communities have brought nearby fringe habitats to the attention of local authorities and other national agencies by opposing encroachment onto land which they consider to be important local resources (see Shoard, 2002 for numerous examples). Nationally, scientific assessments of biodiversity have revealed that it is not those landscapes most revered by artists and ramblers that are richest in wildlife: rather, it is those, at the fringe, that are generally ignored. But we should not imagine that towns are enveloped by rich and diverse habitats; by a jungle of rare plants; by the breeding grounds of newts, warblers and natterjack toads; or by glorious spring-time eruptions of wild flowers. Some contaminated or cultivated sites are as dead as much of the rest of the intensively farmed countryside. But where special habitats or other ecological features have been recognized, there may be clear benefits – aesthetic, recreational or educational – in integrating these with everything else that makes the fringe.

Osment (1998) has argued that for too long, wildlife has been seen as an optional 'add-on' in the planning process. This is clearly changing, but not at any great rate of knots. For Osment, one of the key problems – accentuated by the planning process – has been the ring-fencing of nature: thus, National Parks, Country Parks, SSSIs and other designations have been prioritized and everything else is generally forgotten. There is some evidence to suggest that a more sophisticated understanding of the fringe and its habitats is on the horizon: this may open up opportunities for mainstreaming ecological concerns in models of development and land-use change. Goode (1997), for example, in 'The Nature of Cities' has suggested ways of 'greening' buildings or giving nature a foot-hole in new development. But this is perhaps on a par with the idea that Renzo Piano's 'Shard of Glass' – the unofficial name for the London Bridge Tower – should incorporate a 'winter garden' on each floor. This is perhaps little more than a nod to those who have more serious ecological intent. At the fringe, it is possible to do a lot more than simply acknowledge nature's rightful and essential place in the landscape. Nature can be part of the essential services that ring all built-up areas: it can be incorporated into sustainable drainage schemes of the type found at Howdon and Van Diemen's Land; it might be treated as a partner in flood defence as in the case of the Jubilee River; and it might be used to lighten the footprint of development, including roads, that might otherwise blight the fringe.

The landscape of the fringe is rich in ecological opportunity; the task for planning – an issue that we turn to in the next part of this book – is to ensure that this, and all the other opportunities described in the last five chapters, are not lost either because of planning's heavy-handedness or because of its failure to understand the context of planning at the edge.

Part Three
Planning on the edge

8

Land-use planning and containment

Introduction

Chapters 8 and 9 are closely linked. This chapter examines the traditional 'land-use' model of planning applied at the rural-urban fringe, and also the strategy of containment that has been the key component of edge planning for at least the past fifty years. This strategy is rooted in the UK but has been exported, in various forms, to various other parts of the world. A detailed examination of containment and green belts in this chapter is used to expose some of the problems associated with past planning on the edge – at least in those instances where planning has occurred. This leads into a broader discussion – in Chapter 9 – of the emergent spatial planning agenda and what this may mean at the fringe, and indeed, how the agenda might be taken forward. However, this current chapter is not entirely retrospective, and also examines how current containment policies might continue to promote urban regeneration and prevent sprawl and coalescence, while also evolving into a positive tool for encouraging greener forms of development in the rural-urban fringe. A key argument here – and one that was introduced in Chapter 1 – is that planning at the fringe has not necessarily been *for the fringe*. Containment strategies were designed with objectives in mind that were spatially exogenous to the fringe: protection of the countryside, the prevention of urban coalescence and the promotion of inner urban regeneration. Much of what has happened at the fringe – in areas with or without explicit containment policies – has been organic, and planning has often sought to manage disparate demands for space in an ad hoc way, reacting to development pressure (for example, for the siting of essential services) in a reactive rather than a proactive manner. This is a rather broad statement, and one that is becoming increasingly common in planning literature. Our intention in this chapter is to mould such statements into a more coherent analysis of the failures of planning at the rural-urban fringe. Our broader objective in this part of the book is not to argue for more planning, but to suggest a different approach: perhaps one that departs from conventional planning but becomes more integrated and inclusive. Such an approach – designed to manage the 'multiple functions' of the fringe considered in Part Two – might draw on a softer range of management approaches and avoid emphasis on physical planning; however, it might use existing fringe designations, including green belts, as a framework for future intervention.

Land-use planning and the fringe

The conventional model of land-use planning in England is rooted in the Town and Country Planning Act 1947, which was a product of the post-war reconstruction agenda. The Planning White Paper of 1944 had established that the principal objective of planning was 'the control of land use' (the title of the paper itself), and stated that 'provision for the right use of land, in accordance with a considered policy, is an essential requirement of the government's programme of post war reconstruction' (cited in Cullingworth and Nadin, 1997: 22). The 'right use of land' was to be achieved through a suite of legislation brought forward by successive post-war governments: as well as the Planning Act of 1947, other legislation included the Distribution of Industry Acts, the National Parks and Access to the Countryside Act, the New Towns Act, and a series of Town Development Acts. Together, these established planning as a regulatory framework and as a tool by which local planning authorities (who were invested with the power to grant or deny planning permission following the nationalization of development rights in 1947) could steer development and govern the process of land-use change.

Regulation of development – defined in 1947 as 'the carrying out of building, engineering, mining, or other operations in, on, over or under land, or the making of any material change in the use of any buildings or other land' – was established as the primary function of the system, and this remained unchanged in subsequent Planning Acts brought into statute in 1968, 1971, 1980, 1990 and 2004. However, in the immediate post-war period, the power to regulate was assigned to the counties (as incumbent planning authorities). This system was overhauled in 1968, when boroughs and districts took over responsibility for the development of local plans and the administration of development control, and the role of the counties became more strategic through a framework of Structure Planning. However, at both tiers, emphasis remained on coordinating or regulating land-use change, eventually with reference to a General Permitted Development Order (establishing the actions/developments requiring planning permission) and a Use Classes Order (setting out a classification of land uses and a requirement that 'material changes' from one use to another will require consent from the local planning authority).

This very brief overview of the post-war planning system is intended to underscore the regulatory nature of the system. Many – if not all – of the intricacies have been omitted, yet the overview gives at least a flavour of what the system was about. Cullingworth (1999) has argued that the system was never intended as a merely regulatory tool: it was supposed to be forward-thinking and visionary. However, 'the realities of the post-war years rapidly shattered the early dreams. Other issues dominated the political agenda; plans took much time and effort to bring to an operational level, and the resources available were grossly inadequate to implement them. Both public and private investment was held back' (Cullingworth, 1999: 1). Arguably, the planning system sank into a regulatory mindset following early set-backs. It seemed, at first, to perform this function very effectively because the rate of land-use change nationally was limited and much that was going on in the country – including house building – was public-sector led and therefore easy to influence. In earlier work, Cullingworth (1996: 172–173) has further argued that the increasing failure of the British planning system to plan positively for the challenges of the late twentieth and early twenty-first centuries relates to its inbuilt inability to cope with change. Thus as the assumptions on which the post-war planning system was founded – modest economic growth balanced between the regions, low population growth, little migration (either internally or from abroad), stable patterns of household formulation, stable (low) patterns of home ownership, and slow technological advance – have proved incorrect, debate has increasingly focused on improving the efficiency of the planning system, rather than on questioning its underlying assumptions and processes.

Throughout the past thirty years, there have been periodic attempts to revitalize planning by switching the emphasis between regional, county and development planning, and hence the balance between strategic thinking and local regulation. New legislation has been introduced to modify plans; to allow planning to take account of a broader range of material considerations (allowing it to enter the realm of 'social town planning': see Greed, 1999); to enable it to work more effectively with the private and voluntary sectors; and finally, to ensure that it engages communities in the planning process. However, there was growing consensus during the 1990s that none of these measures were sufficient in the face of the system's intrinsic structural weaknesses. Post-war planning was supposed to be visionary; it was supposed to part of a wider package of social welfare; and it was supposed to engage the general populace in a process leading to broader social, environmental and economic improvement. But a system that empowered only the public sector, that was inward looking, and that failed to concern itself with anything but the control of land, was failing to deliver on these various fronts. In 1999, the Urban Task Force (UTF) called for a more creative planning system, more committed to making things happen and with an overall emphasis on achieving positive change. This, they argued, 'must be based on partnership between the local authority and the project stakeholders, with the full involvement of the local community wherever possible' (Urban Task Force, 1999: 191). The Task Force was largely concerned with the nuts and bolts of planning, though the problems they highlighted reflected broader concerns over what planning should become.

The views of the Task Force also reflected certain undercurrents pushing planning away from its regulatory roots (though the essential regulatory function – the control of development – will always remain) and towards a different model, where planning might become less of a contingent response

– reacting to development pressure – and more of an enabling force, able to orchestrate and coordinate change across different sectors and fields of activity. Before considering the changes (in Chapter 9) that the English system – and indeed wider European planning – is currently experiencing, it is important to reflect on how the conventional land-use model – described above – has impacted on the fringe, especially through green belt policy. Again, these impacts are as a much a result of indirect action, intended to achieve broader objectives away from the fringe, as they are direct intervention.

Containment and the fringe

For half a century, green belts have been instrumental in shaping planning decisions and development around many English cities and towns. They impact on all aspects of land use: on physical development (including housing, industry, utilities and so forth), on agriculture and on the use and potential of the countryside around towns in terms of its capacity to accommodate a range of different recreational and leisure activities. Green belts have undoubtedly affected not only the land they cover, but also the towns and cities they surround and the landscape beyond their area of designation. Their effect on development is often profound. Hoggart, for example, has noted that 'green belt zones provide the land use planning system with its most stringent restraints on new housing' (2003: 158). Indeed, the perceived effect of green belts on housing supply, house prices, and economic growth potential – particularly in south east England – is frequently seen as a case for either amending green belt boundaries (an argument forwarded in PPG 3 [2000]) or doing away with them completely in some instances (Whitehead, 2003).

References to planning and to policy intervention in the opening chapter of this book suggested that regulation in the fringe has tended – in very general terms – to be looser than in the broader countryside (where land-use change is strictly policed) and in cities (where a range of planning and regeneration strategies have been evident and where a certain 'institutional thickness' exists, with a multitude of agencies involved in reshaping and redesigning urban space). But at the fringe, the landscape may appear organic and anarchic (see Chapters 1 and 2), with planning having had apparently less influence, planning controls being flouted, or development control and enforcement turning an apparently blind eye to contraventions of planning regulations. For society at large, and for planning in particular, the fringe can sometimes appear to be out of sight, and out of mind. But in other instances, the existence of statutory green belts suggests that planning has more of a direct, hands-on, concern for what happens at the fringe: containment of this type may create an apparently more regulated landscape, reinforcing the division between town and country.

The original concept of clear physical boundaries between large cities and their surrounding countryside dates back to the late nineteenth century when Howard introduced his proposals for managing growth in London and maximizing accessibility to adjacent green land. His ideas were further developed by Raymond Unwin in the 1930s and by Patrick Abercrombie in the Greater London Plan of 1945. Circular 42/55 (1955) – introduced by Duncan Sandys and the Conservative Government – called upon local authorities to consider formal designation of green belts where it was desirable to restrict urban growth (Elson, 1986: 3–15). Today, green belts cover one and a half million hectares (13%) of England and the policy is regarded by many as one of planning's greatest achievements. According to the latest policy statement (the revised PPG 2, published in 2001) the purposes of the green belt Policy in England are: to check the unrestricted sprawl of large built-up areas; to prevent neighbouring towns from merging into one another; to assist in safeguarding the countryside from encroachment; to preserve the setting and special character of historic towns; and to assist in urban regeneration, by encouraging the recycling of derelict and other urban land (DETR, 2001b: Para. 1.5) Put simply, green belts are viewed as a generic intervention, designed to achieve wider objectives, without a specific purpose tied to the area of designation itself.

The last major study into the effectiveness of green belts concluded that the policy was successful in checking unrestricted sprawl and preventing towns from merging. However, a growing number of voices have questioned the broader value of the policy. Whitehead suggests that London's Green Belt should be scrapped so policy makers can 'concentrate on what is worth saving and use what is not appropriately' (2003: 27). Her statement raises questions about the quality of some of the protected green belt land. Currently all land within designated green belt areas enjoys the same protection, but as some commentators have pointed out, some of it is of little amenity value in itself: 'some is derelict and most is intensively farmed at considerable expense to the taxpayer, while the public has no general rights of access' (Smith, 2001: 7). Bovill has argued that the green belt policy should be kept under review like other planning policies: 'such a review process would probably result in a reduction in the quantity of green belt land with a consequent increase in the quality of the land remaining' (Bovill, 2002: 12). A common criticism of green belt policy has been that the designations are too rigid and permanent and that a more flexible approach is needed. Ron Tate, convenor of the Royal Town Planning Institute's planning policy panel (and the Institute's President in 2005), has suggested that 'we are stuck in a time warp, with the assumption that Green Belts have a life of their own regardless of the planning context' (Dewar, 2002: 8). Since 1997, when 'green wedge' and 'strategic gap' policies first appeared in the revised PPG 7, some commentators and policy makers have been attracted by a less permanent and rigid alternative to green belts. The aims of wedges and gaps are roughly the same as green belts, though they do not aim to assist in urban regeneration efforts or check unrestricted sprawl. The critical difference between these mechanisms and traditional green belts is that they are time-limited and subject to review within the same cycle as other development plan policies. In the future, proposals for additional wedges and gaps would be brought forward in the context of Local Development Frameworks, which are a product of the government's intended 'step change' in planning, brought about by the Planning and Compulsory Purchase Act 2004.

England's green belts – a key planning mechanism affecting the rural-urban fringe – is closely identified with the traditional land-use planning model. Albrechts (2004: 744) notes that this model is:

> basically concerned – in an integrated and qualitative way – with the location, intensity, form, amount, and harmonization of land development required for the various space-using functions: housing, industry, recreation, transport, education, nature, agriculture, cultural activities ... in this way a land-use plan embodies a proposal as to how land should be used – in accordance with a considered policy – as expansion and restructuring proceed into the future.

The green belt is a 'considered policy' in the sense employed by Albrechts (albeit one which both Whitehead and Tate might argue does not adequately reflect local circumstances) administered by central government. However, the extent to which it is integrated and qualitative – or promotes harmonization – is a topic of considerable debate (see Royal Town Planning Institute, 2002; and TCPA, 2002). Arguably the green belt promotes separation rather than integration, though it might still be viewed as a 'regulatory zoning instrument' (Albrechts, 2004: 744) and part of the machinery of land-use planning. And although the land-use model may be broadly concerned with integration and harmonization between land-uses, and between uses and location, often this has not been achieved in the rural-urban fringe, even in the absence of a statutory green belt. This apparent failure has two root causes: first, green belt policy does not have specific objectives within the fringe; and secondly, land-use planning – of which the green belt is a part – has tended to react to development pressure and remained largely unconcerned with forward thinking. The rationale behind much land-use planning has been the need to keep different uses apart: to keep people away from most types of employment uses; to earmark High Streets for retail and suburbs for housing; and to keep everything away from agriculture. Conventional planning has been about designating, compartmentalizing and separating: forward planning has articulated this rationale in proposal plans; and development control has followed it through. But a rationale does not constitute a vision or even a management strategy. This is a broad comment on the planning system, but has particular relevance at the fringe. Throughout Part Two of this book we cited instances where policy intervention might be capable of alleviating particular pressures or addressing concerns outside the remit of land-use planning: the failure to recognize the heritage of fringe areas (Chapter 3); persistent misrepresentation leading to an undervaluing of the fringe (Chapter 4); a failure to ensure that farming makes best use of locational advantages, or to manage the spaces shared by new and surviving economic activities at the fringe (Chapter 5); the need to promote greener waste management strategies (also Chapter 5); and a failure to integrate recreational opportunities with other functions (Chapter 6) or to acknowledge the ecological vitality of the rural urban-fringe (Chapter 7). All such concerns might be addressed within a policy framework and approach – perhaps superseding the green belt – that might be applied more generally to areas of fringe.

Beyond the green belt

In recent years, calls for reform of green belt policy in Britain have become conflated within a broader debate on land and housing supply, especially in south-east England. Edwards, for example, has argued that since the 1980s, inadequate land supply and real growth of incomes for about half of the population in London has resulted in relative house price inflation (M. Edwards, 2000: 108–109). Due to steeply rising property prices and a growing population, more people are now excluded from the housing market and the Metropolitan green belt is under particularly strong development pressure (though other towns and cities such as Cambridge, Oxford and Edinburgh experience similar problems, albeit on a lesser scale). Some of the government's proposed new housing in the South East (ODPM, 2003a) will be on green belt land, but the overall policy may well remain intact and the government has set a target to maintain or increase green belt land in each of England's regions. This means that when green belt land is built on, new land will be added to the green belt elsewhere. Some see this as creeping reform: Dewar (2002) cites the view of the Campaign to Protect Rural England (CPRE): 'the fact that the government has waived through 120 green belt referrals in the last five years shows that green belt policy needs to be strengthened rather than weakened' (Dewar, 2002: 8). Also, those in favour of moderate reform have criticized the government for redrawing the boundaries. Elson has called it a 'take and give green belt' (Elson, 2003: 104) and Fyson, who believes that the green belt is 'grossly overextended', thinks that the inner boundaries have to be 'maintained with some tenacity' (Fyson, 2002: 10). But most green belt developments approved under the Labour Government (in power since 1997) have been of a very small scale (Baker, 2000: 11) and are not part of the radical reform some would like to see. Indeed, some commentators have called for a fundamental rethink of the green belt approach: 'put brutally, is the green belt sacrosanct? However aesthetically desirable it is for London to have a bucolic doughnut, is it rational or fair in a part of the country with such high housing costs?' (Travers, 2002: 11). However, defenders of the policy argue that high house prices are not necessarily the result of a housing shortage. The former Conservative environment secretary, John Gummer, believes that 'rising house prices are much more a reflection of an all-time low in interest rates, and the confidence that they will remain low, than they are of homes shortage'. And even if there is a housing shortage, it does not justify the release of green belt land because 'the incentive to find ways of using brownfield sites will disappear and developers will take the easy option' (Gummer, 2002: 9).

On 'sustainability' grounds it has also been argued that we need to increase development density in already built-up areas rather than allowing further expansions onto green belt land: 'surely there is a better way of dealing with housing problems than tearing up the precious Green Belt and plonking down rows of anonymous houses? . . . Britain is among the most densely-populated countries in Europe yet its urban areas are among the least dense' (Kampfner, 2003: 12). The risk, however, is that new developments do not end up on brown-field sites within the inner boundaries of the green belt, but beyond the outer boundaries, and that too has environmental implications. Leedale (2001), for instance, has argued that restraint policies 'need fundamental reappraisal' because instead of reducing car-based travel they have the opposite effect: 'ironically, and because these constraints are applied far too widely, they force housing development away from infrastructure and accentuate car-based travel' (ibid.: 8).

A critique from more reformist voices has been that only particular interest groups benefit from the green belt policy: 'what was meant to be an instrument to manage change has been captured by particular groups' according to Elson (cited in Dewar, 2002: 8). These groups, as Edwards has suggested, may include 'country landowners, NIMBY residents, and the Council for the Protection of Rural England [now the Campaign to Protect Rural England], allied with green groups, including even quite radical greens, with the formal political parties, all lined up in an unthinking or perhaps opportunistic unison' (M. Edwards, 2000: 109–110). Any radical reform of the green belt policy is unlikely as long as the current political consensus in favour of keeping it is intact. Even the market-orientated Thatcher governments, which deregulated substantial parts of the planning system, defended the green belt. Ambrose has suggested that 'one powerful element in this defence has been the Tory support in the shire counties and outer suburbs for a heavily restrictive approach to land release policies' (Ambrose, 1986: 200–201), but as Morris has argued, defence of the green belt policy 'is not just down to NIMBYism. It's emotional and sentimental. There are few politicians brave enough to face the challenge' (Morris, 2001: 3). In a more conceptual analysis, Rydin and Myerson argue that support for the policy has little to do with whether it is successful or not. Instead, public support for the policy stems from its association with the countryside, which is seen as quintessentially English (again, see Chapter 4) and unlike the city, a healthy and socially cohesive unit. Continuity in the 'green belt concept is not just a matter of continuity in centre-local relations or interest group coalitions but also of more general continuity in cultural discourse' (Rydin and Myerson, 1989: 477). At the present time – and given the wider debates surrounding brown-field recycling and green-field protection – Tewdwr-Jones has suggested that support for green belt policies might be viewed simply as an expression of wider support for the protection of green-field sites.

> However, the pressure for green belt designation is emanating from the combined push of environmentalists and local political representatives, and one wonders whether their call for 'green belts' to be established is in actuality a call for stronger protection of green-field sites. The green belt terminology is a widely recognized form of planning control with the public, and the greater protection of the urban fringe could be more strictly enforced through local authorities'

development plan policies without the necessity to designate a stronger form of restraint.

(Tewdwr-Jones, 1997: 76)

Current debate surrounding green belts has concentrated on the reform agenda, underpinned by two key areas of argument: first, a questioning of the role and purpose of green belts; and second, an insistence by many commentators that there is only a flimsy case for affording such statutory protection to the rural-urban fringe, compared to a much stronger case for accommodating future housing growth and wider development needs. These are the primary concerns at the present time.

However, there are further questions that might be raised that have been largely neglected within the recent literature: the first relates to the way in which green belt designations affect the wider mix of land uses within the rural-urban fringe or might restrict the movement away from traditional land uses to modern re-use (i.e. conversions of the types discussed in Chapter 3); the second concerns the relationship between agriculture, forestry and green belt designations. The green belt has afforded protection to agricultural activities in the rural-urban fringe, preventing the encroachment of physical and 'non-compatible' development. In Chapter 1 we noted that agriculture in the fringe enjoys the advantage of proximity to urban markets, which is particularly useful when local producers employ direct marketing strategies (though the sector's ability to connect with urban markets was questioned in Chapter 5). And thirdly, there are also important questions as to how green belts affect the communities they encapsulate (though such communities may exist beyond the fringe, where green belts extend outwards for considerable distances). The first two of these questions are picked up again later in this chapter, where we begin to consider the role of the green belt, and land-use planning more generally, in managing the landscape of the fringe. But the questions need also to be placed in the context of ongoing debates over alternative containment strategies and the current rethinking of the green belt being led by a range of groups in the UK. Might alternative 'containment' strategies for the fringe be better placed for dealing with planning and management concerns within the fringe? And might this herald the beginning of planning *at* the fringe, *for* the fringe?

Alternative strategies: wedges and gaps

There is already some indication that planning should adopt a more flexible approach to the rural-urban fringe. Planning Policy Statement 7 (ODPM, 2004c) states that:

> While the policies in PPG2 continue to apply in green belts, local planning authorities should ensure that planning policies in LDDs address the particular land use issues and opportunities to be found in the countryside around all urban areas, recognising its importance to those who live or work there, and also in providing the nearest and

most accessible countryside to urban residents. Planning authorities should aim to secure environmental improvements and maximise a range of beneficial uses of this land, whilst reducing potential conflicts between neighbouring land-uses. This should include improvement of public access (e.g. through support for country parks and community forests) and facilitating the provision of appropriate sport and recreation facilities.

(ODPM, 2004c: Para. 26)

The policy statement also tries to create a less rigid context for farming in the fringe, by calling on authorities to give consideration to all and any (well conceived) applications to diversify into non-farming enterprises (ibid.: Para. 30iii). This could mean that some of the barriers to diversification in the fringe (see Chapter 5) might eventually be removed. But the tone of this and other recent policy edicts is rather odd: the government recognizes that planning has been restrictive at the fringe (where the green belt exists); but it is now saying that although containment will survive, it should be possible for planning authorities to reflect development needs. There are clear contradictions here. If the government intends to pursue the objective of containment then it will need to harden its stance towards development in areas of green belt. If, on the other hand, it wishes to promote certain types of activity, then a softer approach to containment will be required: something will have to give. This might mean moving to a different 'management' mechanism, which does not have containment or protection as its primary function, but which prioritizes the 'special needs' of land within the rural-urban fringe.

In 2002, the ODPM published a study into the potential and possible benefits of local 'containment designations' including 'strategic gaps', 'rural buffers' and 'green wedges'. The new PPS 7 (Sustainable Development in Rural Areas) makes no reference to these designations though they might feature in other forthcoming Policy Statements (only 7 have been published to date, and 18 are outstanding, assuming that each of the PPGs will be replaced with its own PPS). However, the older PPG 7 stated that when reviewing local plans, authorities should reassess the function and justification of local containment mechanisms, perhaps examining their broader objectives. Responding to a House of Commons Environment Select Committee Report on Housing, which touched upon the issue of local designations, the government puts its support behind local designations that perform the same function as green belts and have the same objectives and purposes. Two points are clear: first, the government is keen to see a more flexible approach to planning (and to management?) at the rural-urban fringe, even where green belt policies apply; and second, there is some willingness to look beyond statutory green belt, and towards local designations.

The aim of the ODPM study (2002c) referenced above was to assess the current role and status of local policy designations and to examine alternative options for the future. In part, the remit was to assess the justifications for including particular land areas in such designations; to establish the objectives for land use within such areas; and to gauge the extent to which these fulfil

Alternative strategies: wedges and gaps 171

current green belt policy purposes as set out in PPG 2 (which is now set to be revised). The purpose of each of the three types of policy designation – gaps, buffers and wedges – was identified as follows:

- 'Strategic gaps'; they are to protect the setting and separate identity of settlements, and to avoid coalescence; to retain the existing settlement pattern by maintaining the openness of the land; and to retain the physical and psychological benefits of having open land near to where people live;
- 'Rural buffers'; to avoid coalescence with settlements (including villages) near a town until the long-term direction of growth is decided; and
- 'Green wedges'; to protect strategic open land, helping to shape urban growth as it progresses; to preserve and enhance links between urban areas and the countryside; and to facilitate the positive management of land.

Some overlap with green belt functions was clearly identified within the ODPM study, but there are a number of important departures. First, buffers and strategic gaps aim to prevent the coalescence of villages, whereas green belt policy focuses explicitly on larger towns. The new tools might therefore address a wider range of rural-urban fringe settings. Second, within buffers, gaps and wedges, land is protected because of its own intrinsic qualities, and not merely because of its proximity to a town. Therefore, the new tools present an answer to the criticism that green belt policy is blind to land-quality issues and merely conserves for conservation's sake: the new local tools are based on a strategy of appropriate protection. They edge towards the aim of ensuring planning in the fringe, for the fringe; or at least recognizing that the fringe has some intrinsic qualities worth managing. And third, green wedges penetrate into towns and cities and form links between urban open space and outlying countryside. Again, this suggests that the fringe is valued as a resource in its own right.

Wedges are thought of as management tools – there to be used flexibly – and not as restrictive girdles enveloping towns and cities. So while the green belt has tended to promote development 'leap-frogging' beyond the designated areas, Green Wedges can be used to promote development 'fingers', stretching out from the built-up area, where these are deemed appropriate. The views of local authorities were highlighted in the ODPM study: many considered wedges, gaps and buffers to be an improvement on green belts, allowing authorities to aspire to and achieve wider objectives (of the type emphasized more recently in PPS 7, Para. 26) within the fringe. The government wishes to see a more flexible approach to the management of recreation in the fringe, and a recognition that people work in fringe areas (it has an economic function). But the government's own aspirations, again set out in PPS 7, hit the barrier of its own green belt policy. In this context, the ODPM study (despite predating the new PPS 7) brings out a number of important messages in relation to containment and land-use planning and management in the fringe:

172 *Land-use planning and containment*

- Locally developed restraint policies may offer a positive means of managing different forms of development in the rural-urban fringe; local priorities can be more easily accommodated;
- In particular, green wedges penetrate into towns and cites and form links between urban open space and the countryside; they address the needs of fringe areas that extend inwards from the countryside around towns;
- These alternative restraint policies have numerous advantages over green belts including the flexibility to allow release of land for development in sustainable locations; and they also allow local authorities to make judgements over the appropriateness of landscape protection in different locations;
- However, they are not as permanent as green belts and there is not the same certainty in the presumption against certain types of development; this lower level of permanence could lead to the creation of 'hope value' for landowners and speculative development pressures.

The final point is often emphasized in policy debates: end the certainty provided by green belts, and there is a risk of opening up the fringe to urban sprawl (Campaign to Protect Rural England, 2005: 4). We return to this point below. However, how can local authorities engage fully with economic development, with landscape management, with issues of building re-use, and with recreational development if their hands are tied by restrictive national polices? This question is the focus of current debate. It is argued that locally designated policies may be more flexible in their design and application: such tools may also be more sensitive to changes in land-use demand, permitting necessary diversification or essential use-changes. They certainly do not preclude peripheral development (including urban extension in the form of development fingers) and are therefore more consistent with the objectives of housing policy (and wider government thinking on sustainable urban form). Putting this in the context of our Chapter 1 discussion, local tools may give authorities the opportunity to at least think about 'multifunctionality' and the integration of land uses, and to start addressing some of the conflicts (e.g. between farming and more urban uses – see Chapter 5) that have developed over the past 60 years. Local mechanisms – incorporating a management approach – may enable authorities to counter the argument that containment does not provide a positive agenda for the rural-urban fringe. But supporters of green belts, including the Campaign to Protect Rural England (CPRE), marked 50 years of the policy in 2005 (i.e. 50 years after the issuing of Circular 42/55) with a celebration of its success, arguing that it 'is the envy of other countries and remains one of the sharpest tools in the planning toolkit, maintaining the openness of the countryside around the English towns and cities it embraces' (CPRE, 2005: no page number). In an earlier statement, the CPRE expressed support for local designations, but argued that such designations are less satisfactory in many instances than a formal green belt (CPRE, 2001). The green belt remains, in the view of the CPRE, the most powerful policy tool available for aiding urban regeneration and protecting the countryside that any local authority can possess. However, these are clearly not the only considerations within the rural-urban fringe; many fringe areas have a wider range of needs (and functions, as noted in earlier chapters) and mere protection is not a sufficient means of promoting different uses or balancing a range of priorities. The CPRE cites the fact that green belts can be found within 'some of the most prosperous parts of the country' (CPRE, 2005: 8) as evidence that it does not stifle economic development. But can the fringe engage with the new economy described in Chapter 5 without greater flexibility, and can it continue to perform those service functions necessary to support economic growth? Furthermore, is it the best framework in which to secure landscape improvement or promote ecological diversity? These are some of the questions recently addressed in various reviews of green belt policy, all aiming to contribute to this debate in the run-up to the expected publication of PPS 2.

Green belts: Is re-invention possible?

May 2002 saw the publication of policy statements from three different bodies: the Royal Town Planning Institute (RTPI), the Country Land and Business Association (CLA) and the Town and Country Planning Association (TCPA). The RTPI published a statement on 'Modernising green belts' (RTPI, 2002) which followed an earlier discussion paper (RTPI, 2000). Key problems identified with the policy (cited in O'Neill, 2003) included:

- A failure to keep pace with other aspects of the changing planning policy agenda of recent years (mirroring the broader criticisms of land-use planning articulated by Cullingworth, 1996), including the 'spatial planning agenda' and the desire to move planning away from its retrenchment into land-use control;
- The public perception of the role and purpose of green belts being out of step with reality, including the need for urban extensions, the promotion of Growth Areas, or the need to upgrade infrastructure required to support economic growth;
- The conflicting aims and objectives in the application of green belt policy, which have perhaps become more acute since the issuing of PPS 7 in 2005.

The RTPI argued that particular problems arise from implementing the sustainable development agenda. For example, cutting down the need to travel, the use of brown-field land, and meeting housing needs in optimal locations are often at odds with the strict prohibitions of green belt policy at the rural-urban fringe. It is accepted that green belt policy is probably the planning tool best known by the general public, but it is also the least understood. Rather than being a tool to protect countryside (the view held by the CPRE) it is a regional strategic planning tool, which needs to connect with a wider sustainability framework, and with economic and regeneration objectives. This means that green belts should perform a central role in effective spatial planning from the strategic to the local level; and that policy should be expressed in terms of the key principles to be satisfied, if green belts are to be approved in development plans (again, this might mean incorporating the objectives set out in PPS 7). But the present guidance in PPG 2 presumes against development, except under exceptional circumstances, within designated green belts. Contrary to the view held by the CPRE, the RTPI has argued that this can have negative economic and environmental consequences. For example, farming in the rural-urban fringe – and within green belts – has only marginal viability (according to the RTPI, and this is despite proximity to urban markets – see Chapter 5) and may be one of the first activities to be abandoned in fringe locations. Green belt designations place significant obstacles in the path of economic and land-use diversification; this means that as farming activity declines, it is difficult to secure replacement activity and land formerly under agriculture may be abandoned and lose much of its previous 'quality'. It follows that green belt designation can drive, or at least accentuate, the degradation of rural-urban fringe land, contributing to economic abandonment. This is unacceptable, according to the RTPI, in those areas where urban populations look for fresh air and recreation: it is suggested that:

> There must be a much more pragmatic approach to the control of development in green belts, with a lot more flexibility and realism than is displayed in the current guidance [and] policies must provide scope for economic diversification, so that funding is available for environmental management and enhancement, and green belts are not reduced to museums of inactivity.
> (RTPI, 2002: 7)

The RTPI makes the link between the rigidity of development control and the capacity of authorities to implement environmental enhancement and management programmes in the green belt: policies that largely preclude development make it more difficult to secure adequate funding for enhancement programmes, and policy may well obstruct the 'sensible modernization' or redevelopment of even small sites. The CPRE counters this view by arguing that:

> Councils already have powers [under s215 of the Town and Country Planning Act 1990] to require the amenity of neglected land to be restored . . . positive management is the answer to neglected land; permitting its development serves only to encourage landowners to foster neglect in order to cash in on a hefty development premium.
> (CPRE, 2005: 8)

Of course, it might be difficult to get a bankrupt smallholder to restore amenity value. The RTPI argues that it is vital that there be mechanisms in place – perhaps written into 'containment frameworks' to secure implementation of positive policies to tackle fringe problems and to ensure environmental improvement, and so maximize the amenity value of green belts. Such measures are sufficiently important to justify making the approval of green belt designations conditional on local authorities, and their partners, bringing forward positive strategies and programmes for environmental improvement, farming and forestry, leisure and rural settlement support as well as proper provision for urban-related infrastructure and utilities. Clearly, there is a view that the framework should carry a broader range of policy objectives, many of which should be particular to the needs of specific areas. The green belt might be subject to an 'enhancement' that focuses less on protection, and more on promoting green development and integrating the needs of the multiple fringes examined earlier in this book. A programme aiming to 'modernize' green belt policy might consider a number of issues (RTPI, 2002: 10): extent and shape; time horizon; environmental quality; development control policies; farming and forestry; and longer-term development opportunities (see Table 8.1).

Given our analysis of competing pressures and conflicts at the fringe, a strategy of revisiting green belt policy in the context of a spatial planning agenda (see Chapter 9) seems sensible. green belts and PPG 2 are already out of step with revisions to planning guidance contained in the emergent Planning Policy Statements, and especially PPS 7. That said, it will not be easy to reach agreement on the future form of green belt policy. When reformers call for flexibility, the environmental lobby derides this as weakness; similarly,

Table 8.1 Green belt: an agenda for modernization (RTPI, 2002)

Extent and shape	*Environmental quality*
The broad extent of green belts should be drawn up and reviewed at the regional level in the context of the regional sustainability framework. The extent will be shown only in diagrammatic form, and will be determined by the settlement policy that reflects regional regeneration and economic development objectives. A green belt need not fully surround a settlement in order to serve its purpose. Over-arching policies such as reducing the need for travel will influence the shape and extent of the green belt. The relationship of the boundary to urban form will be an important consideration. There should be scope to renegotiate the configuration of the urban edge where this will benefit the sustainability of planned development. And boundaries must be related to physical features whether natural or man-made, to give an obvious edge that is clear on the ground as well as on the map.	Policy needs to pay more attention to the environmental quality of land in green belts. They are the most accessible open land for many urban dwellers, to which they look for fresh air and recreation. Yet almost all green belts contain traditional rural-urban fringe activities and substantial areas of land that is environmentally degraded. This issue might be tackled effectively if the approval of a green belt was made conditional on bringing forward policies and programmes of environmental management and enhancement to tackle traditional problems of the fringe. For example: development of a green network linking town and country, access agreements with landlords to facilitate informal recreation, and financial contributions from nearby approved developers.

Table 8.1 continued

The time horizon
If green belts are an integral part of the spatial strategy then logically the green belt boundary should remain valid for no longer than the time horizon of the plan: if the plan has a ten-year time horizon, so has the green belt. The green belt then ceases to be sacrosanct and becomes integrated into a wider evolving strategy. Some alterations of boundaries are probable in response to changes within a plan, monitor, and manage regime.

Development control policies
A more positive and flexible approach is needed. A more relaxed attitude to development in the countryside in general needs to be reflected in green belts in particular since there is no logic in more restrictive policies in green belts than those in adjoining areas of countryside. There needs to be sufficient development value in green belts to fund their environmental enhancement. Limited development might be permitted provided appropriate uses, siting, and landscaping are fully addressed. National policy guidance could be designed to leave room for local discretion and flexibility.

Farming and forestry
There is a need for explicit strategies for farming and forestry within green belts. The RTPI suggests that farming may have a limited future in the fringe. However, it acknowledges that the strategic approach seen with community forests demonstrates effective green belt application, with close links to sustaining and enhancing the environmental quality of green belts.

Longer-term development opportunities
There is an issue about the way in which longer-term development opportunities extending beyond the time horizon of the plan are handled. The RTPI considers it important that particular areas of land are not given an interim status. Where the time horizon of a green belt designation is limited it is preferable to place reliance on regular and timely plan review, rather than making early but tentative provision for possible long-term development needs.

the responsiveness lauded by the RTPI is viewed by some as simply a concession to development. But we have to return to our key question: what do green belts really achieve within the rural-urban fringe? The Town and Country Planning Association (TCPA) – a proponent of the original policy – has put its weight behind calls for a review of green belts. Its 'Policy Statement on Green Belts' (2002) argues that belts in south east England have become little more than 'green blankets' extending 20 to 30 miles into and beyond the fringe. Their role, according to the TCPA, should be refocused and bent towards achieving sustainable development. Again, the view that green belts should 'do something' finds support. In 2002, the Association argued that the policy lacks imagination and its Chairman at the time argued that a more radical vision would see areas of green belt becoming 'eco-parks, accommodating farming and horticulture that can help feed neighbouring towns and cities, encourage farmers' markets and local sourcing of food.

Green belt should also provide extensive public access, have country parks and recreational space, be rich in wildlife, woodland and unimproved grassland'. This would mark a departure from the status quo in which 'we have dormitory towns and villages embedded in the green belt, unable to attract jobs, or provide homes to support local schools and shops'. This departure would be possible if green belts were 'modernized' and local authorities given more powers to de-allocate land. Like the RTPI, the Association viewed both local control and local designation as key components of future policy, and recommended:

- Reviewing green belt policy to assess how far it is relevant to present and future needs and how it can contribute to the achievement of sustainable development objectives;
- Encouraging the introduction by planning authorities of more limited strategic gap and green wedge policies as a substitute for broad swathes of green belt;
- Joint action with the Countryside Agency (soon to become part of 'Natural England') to make an appraisal of how far development control in green belts hinders the agricultural sector and the strengthening of rural economies, and how far the rules applying to existing development should be revised;
- Encouraging more effective programmes in green belts to achieve the land-use objectives set out in PPG 2 (and now expanded upon in PPS 7).

Perhaps not surprisingly, the CLA has joined the chorus of voices calling for reform, arguing that green belt policy does nothing to assist communities (and economic uses) 'embedded' in areas of designation. It adds that rural-urban fringe businesses and communities are not benefiting from broader countryside policies that address social, economic and environmental needs. Again, this point seems to have been accepted in the new PPS 7. The Association calls for a change in policy that embraces the continuing role of agriculture and horticulture in green belts, including farm diversification; the need for economic and social development to achieve regeneration of the rural – and the fringe – economy; and the need for more housing, where appropriate, to meet local needs. All these critiques of the green belt share the view that a policy aimed primarily at containment and protection is insufficiently positive and forward thinking. We suggested in Chapter 1 that the fringe has often been neglected, and this view seems substantiated in the recent reviews of green belt policy cited earlier. Re-invention – if this is possible – might mean moving to a 'rural-urban fringe management framework' in which gap, buffer and wedge policies were placed on an equal footing with green belt policies, and where all such tools were applied by local authorities – perhaps in partnership with regional bodies (including the Assemblies and the Development Agencies). Critically, the movement from 'containment' to 'management' would increase the scope of frameworks to achieve something more positive, and to empower local authorities to make choices with regard to development, protection, landscape enhancement, diversification and so forth: choices that they have hitherto been unable to make. In the next chapter (9) we consider the principles that might underpin such a policy approach at the rural-urban fringe, linking these with the emergent spatial planning agenda. Clearly, a key aim of any future approach will be to 'manage' and integrate – where possible and desirable – the needs of the different 'fringes' discussed in Chapters 3 to 7.

In terms of the 'historic fringe', future strategies will need to address issues of re-use of the historic legacy (particularly the industrial legacy), and how future recreation and economic development can best make use of the fringe's historic assets. With regard to the 'aesthetic fringe', there is perhaps a need to think beyond physical attributes and work out what people actually value about the landscape of the fringe. The TCPA called for the fringe to become an 'eco-park', but then balanced this view with an apparent nod to the value of 'unimproved' grassland. Management at the fringe will need to be inclusive in the sense that it reflects what communities of users want the fringe to become. The green belt has sought to promote a 1950s ideal of the perfect landscape, resisting all change. But management will become more complex if competing views of what the fringe might become are given a voice. Some will certainly support the 'eco-park' option while others may prefer a less sanitized 'vision' of the future fringe. Any management framework would need to grapple with such issues from the outset, providing local authorities with a clear set of principles for mediating between competing pressures in the fringe (these pressures already exist, but the green belt has glossed over them, to the detriment of some activities including farming).

In terms of the 'economic fringe' (Chapter 5), we have said little about the pressure exerted by green belt policy on general economic development, but have made some references to the problems planning restraint may cause in respect to farming. All the recent reviews of green belt policy agree that future policy should reflect the needs of the farming sector in the fringe. The CLA has been particularly critical of containment policies and their impact on agriculture, and has called for an easing of restrictions. However, the recent revision of PPG 7 (and its transformation into PPS 7 [ODPM, 2004c]) has led to a reappraisal of those economic uses that might be supported in the wider countryside, given that farming is no longer the principal income generator in rural areas. Clearly, this same concern applies in the fringe where a wider range of non-farming enterprises might be encouraged. However, the CLA's primary concern is with the land-based economy, and it wishes to see a review and not a removal of green belt policy. It favours additional funding for agri-environmental and forestry schemes in the fringe: 'green activities'. But other 'green' activities might find a place in the economic fringe, including 'green' energy generation, commercial waste recycling or other 'environmental conscience' activities.

In this chapter we have noted that strict containment policies can be bad for communities within or adjacent to the rural-urban fringe. It may be socially divisive, causing a gentrification of embedded green belt towns and villages. This problem coupled with its restrictions on economic growth and diversification may mean that green belt policy, as currently set out and implemented, reduces social inclusion, and contributes little to community renewal, regional prosperity, sustainable communities, or rural diversification. But in our earlier discussion of the 'sociocultural fringe' (Chapter 6) we suggested that the fringe is largely a 'landscape without communities'. We make this claim because a green belt may extend 30 miles beyond built-up areas, while areas that we consider to be fringe are often much narrower. The fringe and green belt are not synonymous, though the green belt is an important expression of planning policy within the fringe.

Therefore, in terms of the sociocultural function, we have suggested that the fringe has a community of users rather than a community of residents. Future management strategies need to ensure access and facilities where appropriate. The social-cultural fringe is a product of all the other fringes discussed: it is economic in the sense that leisure and recreation are (or could increasingly become) part of the economic life of these areas; it is aesthetic in the sense that landscape becomes something that people value and enjoy through recreation; it is historic because the industrial legacy of the fringe is tapped as a source of visitor attraction; and it is ecological in the sense that people may visit the fringe for health or educational reasons linked to the natural environment. The 'ecological fringe' was examined in the last chapter: there we saw that the fringe has been reinvented in some places as wetland parks (as in the case of Howdon Wetlands and Barn Elms), providing a sanctuary for habitats and wildlife effectively displaced by urban development. Elsewhere, abandonment and neglect have provided the ideal precursors to ecological diversity.

The purpose of the next chapter is not to establish a blueprint for the management or planning of the rural-urban fringe, but given that a fairly detailed picture of the fringe has been established, and that green belt policy is viewed as not providing the best framework for managing the fringe, it seems appropriate to reflect on how a spatial planning agenda might be taken forward in the rural-urban fringe, and what kinds of principles might guide its development.

Conclusions

Griffiths (1994) has argued that there has been little in the way of proactive planning at the fringe; rather, 'conventional' planning has been a passive observer of land-use change, or simply policed containment policies where they exist. We have argued in this chapter that there has been no integrated management of change in the rural-urban fringe, or any real reflection on the role and functioning of these landscapes. This is largely due to local authorities being disempowered at the fringe, and consequently unable to deal with issues of economic development or landscape management. There has been a lack of integrated thinking because authorities have had their hands tied. But that is not to say that the failure of planning at the fringe is all due to national policy. According to Shoard (1999), the lack of integrated thinking in rural planning (and planning more generally) is a feature of the past 60 years and is only now being challenged with the emergence of a sustainable development paradigm and a spatial planning agenda. At the rural-urban fringe, the land-use model – as summarized by Albrechts (2004) at the beginning of this chapter – has not achieved even its own limited objectives: first, because development and strategic planning has generally ignored the issues and concerns specific to fringe areas, and secondly, because the model is passive and localized (ibid.: 745), and as such, fails to bring together those partners with an interest in the fringe (this particular failure is addressed in Chapter 9). Fringes have been recognized to be complex landscapes (Thomas, 1990) and it is perhaps this complexity and dynamism that the land-use model cannot address (Cullingworth, 1996, 1999), though the designation of green belts has meant that local planning has never been called upon to tackle this inherent complexity. Fringes are dynamic interfaces where different interests compete: public or private utilities, expectant land-owners, user communities, wholesalers and distributors, farmers and foresters. On occasions, there has been some coordinated thinking (as in the case of Colliers Moss, examined briefly in Chapter 1), but generally:

> Co-production of plans with the major stakeholders and the involvement of 'weak' groups in the land-use planning process are . . . non-existent. The whole apparatus of adverse bargaining, negotiation, compromise, and deadlock, which normally surrounds the planning process, must be questioned.
>
> (Albrechts, 2004: 745)

Land-use planning, as a passive contingent response to development pressures, has fared particularly badly at the edges of cities. Is it possible that a spatial model, providing an enabling force for stakeholders, could fare better? This question should perhaps be addressed in the context of a specified objective for planning and management at the rural-urban fringe. So far, we have not established any particular objective, but have simply pointed to the diversity of functions and interests that share space in the fringe. Defining what the future fringe should be is clearly a controversial topic: there are those who expound the virtues of cleaning up the peri-urban landscape, and of making it the latest national urban design project. On the other hand, many commentators find much to admire in the unimproved fringe (Shoard, 2002). What the fringe should be, from an aesthetic or design perspective, is not a debate that we wish to engage with. This is reflected in the next chapter (9) which focuses attention on the principles – integration of land uses and activities, interaction between functions, accessibility, and making best use of current and future assets – that might be incorporated into a future fringe planning and management framework. But these do not reflect a belief in the need for *more* intervention in the fringe, or a need to sanitize these hitherto 'lawless' landscapes. They are merely presented as an interpretation of how spatial planning (which explicitly rejects any prioritization of public-sector-led land-use planning) might be adapted to achieve fringe objectives – running with the grain of multifunctionality – without reference to what the aesthetic or design objectives might actually be. We have chosen to leave this debate to the policy makers.

9

Planning reform and the spatial agenda

Introduction

We saw in the last chapter that the land-use planning model and containment strategies have tended to promote a narrow 'vision' for the rural-urban fringe, and rarely reflect the needs of the 'multiple fringes' (or the need to integrate land uses) discussed in Part Two of this book. In this current chapter, we move beyond the status quo to look at the particular opportunities arising from planning reform in England, and the wider 'spatial planning' agenda emerging across Europe. It is not our intention to provide a model or blueprint for planning at the rural-urban fringe. Rather we use the discussion of multiple fringes as a context for understanding the principles that might underpin future policy interventions. We believe that the examination of the historical, aesthetic, economic, sociocultural and ecological functions of fringe landscapes has revealed a need to develop a model of intervention, or management (this is not a comment on the *extent* of intervention required) that is grounded in clear leadership (of the type displayed at Barn Elms); effective partnership working (the hallmark of the Community Forests Programme); integrated management (shared between different agencies with different 'sectoral' interests); and inclusivity (something that has characterized most, if not all, of the project examples cited thus far). And through an approach to planning and management that avoids being reactive – and knows what it wants to achieve in particular fringe areas – intervention might at least explore the possibility of achieving greater integration between land uses and interaction between fringe functions. It will certainly seek to combine recreation with economic development or ecological projects of the type illustrated in Chapter 7. It might also explore more unusual connections, perhaps between energy generation and industrial uses: though it might be possible to do more than simply position solar panels on warehouse roofs. It became clear in Chapter 2 that *people* have been squeezed out of some fringe areas, often because new economic uses and transport infrastructure seem to have an insatiable appetite for land. But in Chapter 6, it became apparent that landscapes often become valued because they are accessible, and provide opportunities for people to pursue different activities, some that might be labelled 'green', and others that might be viewed as more conventional (Chapter 6 explores this distinction at some length).

After introducing the 'transition' to a spatial planning model in the first part of this chapter, we examine the principles noted above in greater depth. This leads into a review of the planning reform programme in England – instigated by the Planning and Compulsory Purchase Act 2004 – and finally into a discussion of current practices and models that might inform future policy interventions, where necessary, in the rural-urban fringe.

Spatial planning at the edge

Land-use planning is predominantly a physical tool: controlling the use of land in an attempt, in some instances, to influence broader social and economic outcomes. Spatial planning is not a departure from the land-use model, but a complementary force. It is also a label for those practices seeking to broaden the remit of planning: planning as an enabling force that mobilizes wider support and resources behind a vision; planning that is not necessarily public-sector driven but built on strategic and local partnerships; and finally, planning that reconciles competing interests and seeks to do more than merely control the use of land. Indeed, Brandt and Vejre's (2003 [2000]) application of 'multi-functionality' to planning, introduced in Chapter 1, fits comfortably within this agenda, being concerned with how places work, and with possible integration across land uses, and interaction between functions.

Drawing on the work of Kunzmann (2000) and Healey (1997), Albrechts suggests that strategic 'spatial planning is a public-sector-led socio-spatial process through which a vision, actions, and means for implementation are produced that shape and frame what a place is and may become' (Albrechts, 2004: 747). Similarly, drawing on the work of Brandt and Vejre (2003 [2000]), 'multifunctionality' may be viewed a framework for understanding places and as a conscious guide for their future planning (Gallent et al., 2004). Tewdwr-Jones (2004: 562–563) suggests that spatial planning has an integrative and coordinative role, is participatory and informative, and concerned with management, distinctiveness, difference and place-making. He adds that it is value-driven (proposing positions and negotiating interests), and action-orientated (seeking to realize proposals through whatever means necessary; ibid.: 562). It might be thought of as an enabling force rather than a regulatory mechanism, though such mechanisms retain importance despite any shift in planning's substantive or conceptual focus (there will always need to be 'development control'). Albrechts (2004: 749–750) echoes these conclusions, but characterizes the move from land-use plans to strategic plans as a transition, with land-use plans seeking to control change, guide growth, promote development, regulate private development, and to engage in technical or legal regulation. In contrast, strategic (spatial) plans are concerned with longer-term visions and short-term actions, with frameworks or guidelines for integrated development, operating through the interests of selected stakeholders, managing change, and representing a negotiated form of governance (ibid.: 748).

As we have noted at various points in this book, but especially in Chapter 8, past planning at the rural-urban fringe has proven inadequate in a number of respects: critically, it has failed to articulate any particular view of the functioning of the fringe, failed to manage change, to promote integrated development or to engage effectively with stakeholders. Rather, the system has either ignored the fringe or sought to prevent potentially beneficial development (either through its presumption against development in 'open countryside' or because of the designation of green belt). In cases where planning has failed (or been unable) to realize the potential benefits of intervention, we can attribute at least some blame to its land-use control focus. But by using the alternative model of 'multifunctional management' suggested by Brandt and Vejre (2003 [2000]), it is possible to build a case for interventions that seek to harmonize farming with local ecology; to allow the landscape to maintain its economic function by permitting farmers to diversify into new activities; to expand sociocultural functionality by developing new access opportunities and new facilities; and to link essential services with landscape enhancements where they are deemed necessary. This is not a strategy for 'sanitizing' the fringe, but for allowing sensible interventions that have not always been possible under the land-use model (which compartmentalizes functions) or areas under statutory green belt (which freezes those functions and threatens their future viability). The detail of possible interventions – and functional interactions – is an issue for local consideration, and one that we do not touch upon in this chapter, though a number of examples emerged in Part Two; some are revisited in Chapter 10.

The model set out in Figure 9.1 emerges as a product of the 'context' for planning examined in Part Two. It is an adapted spatial planning agenda, pointing to the need for leadership and partnership, integration and inclusion, all feeding into 'place-making' at the fringe. It is concerned with understanding what partners – with different priorities – wish to achieve and offers a very simple model for action: it focuses on partnership, integration and inclusion because of our recognition that there are 'multiple fringes' that need to be managed in unison if any coherent outcomes are to be achieved. This is obviously a very general statement on how planning might deal with the fringe, but if put into action, it would mark a considerable departure from past practice. As we noted earlier, land-use planning has sought to keep functions apart; it has been largely public-sector led; it has often paid little more than lip-service to the need for community participation and hence inclusion; and it has rejected the model of 'negotiated governance' integral to spatial planning and to many of the successful fringe projects catalogued in this book. What we are really talking about is a mainstreaming of many of the practices explored in our case examples, but situated within a spatial planning model. The Countryside Agency has lauded the achievements of the Community Forests Programme: we would agree that the Partnership Forums provided clear leadership (on most but not all objectives); that the partnerships themselves were often well constructed; and that communities became involved at

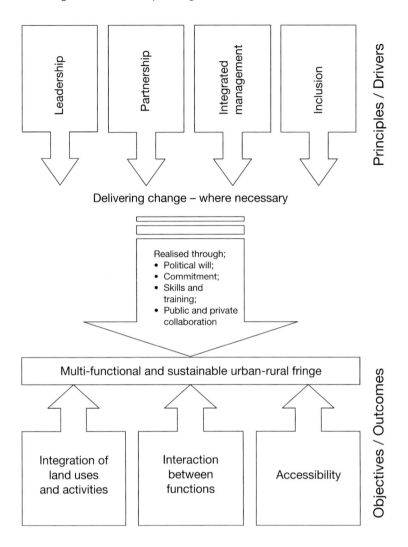

9.1 Principles and outcomes – spatial planning at the rural-urban fringe.

the grass roots. But the partnerships were public-sector-dominated and found it difficult to work with landowners or private enterprise. The vision of 'environmental regeneration' was successfully sold to local government and to communities, but private sector interest never really took off. The private sector only really became involved where development permissions were sought and planning authorities' negotiated agreements aimed at extracting funds to further regeneration objectives (see Counsell, 1995; Oxford Brookes University, 2002). The Community Forest Programme had (and retains) a certain vision of the 'countryside around towns': bucolic, attractive, regenerated, clean, safe, and a magnet for people and for investment.

We have no such 'vision' in mind, but if it is necessary to talk in terms of objectives, we would point to the fringe's historic legacy as something worth valuing (Chapter 3); to debates over what constitutes an attractive landscape (just the foliage of the Community Forests, or something more diverse? Chapter 4); to the varied economy of the fringe, which is surely worth supporting (Chapter 5); to its existing and possible future socio-economic functions (Chapter 6); and also to its ecology, of which woodland is but a small part (Chapter 7). As noted at the end of Chapter 8, we have no single objective for planning at the edge, other than to see some integration of and interaction between these functions; and also, of course, to improve access that will ensure that the opportunities linked to these five functions are shared equitably. This is a 'vision' of a kind and will only be realized through political will (an acceptance that the fringe has a potential worth exploring); commitment (to working differently, and to committing resources to the fringe); skills and training (which enable people to work differently); and through more effective collaboration between private and public interests: we touch upon these issues, very briefly, in Chapter 10. We saw in Chapter 7, in particular, that public–private partnerships have been especially weak at the fringe, possibly because landowners choose to do nothing with landholdings in the rural-urban fringe, and simply live in the hope that they will one day secure planning permission that will bring a windfall profit. This view is tied to the land-use planning system which, as far as landowners are concerned, simply creates winners and losers. But what if a redesigned green belt created the conditions for certain types of development in the fringe, and gave local communities a stronger voice in determining the form this development might take? Landowners might no longer live in the hope that, one day, their ship would come in and they could despoil the landscape with yet another appalling red-brick housing estate. Rather, they would know that development opportunities existed, but only ones in keeping with the local authority's (together with the nearby community's) management framework for the fringe, perhaps as a 'multifunctional', ecologically creative landscape.

The principles set out in Figure 9.1 are expanded upon in Table 9.1. All four principles – leadership, partnership, integrated management and inclusion – add up to a model of 'negotiated governance'. This is intended to shape thinking on the rural-urban fringe, producing an outcome that is distinct from the norms of land-use planning. The land-use model has been inert at the fringe, often being passive in the face of development pressure, except in those instances where statutory containment instruments have insisted on tighter regulation. It is possible to be more creative – as a number of examples have demonstrated in Chapters 2 to 7 – particularly in terms of how multi-functionality is achieved.

The principles explored in Table 9.1 provide an approximate guide for action. The intention now is to reconsider these principles in the context of planning reform that should, in theory, address the areas of leadership (its form), partnership (its operation), management (its structure) and inclusion (its processes). We have already noted that the Community Forests Programme has been held up as a model – good enough for potential mainstreaming – for

Table 9.1 Principles for policy intervention

Leadership
Leadership is important both in terms of pushing the fringe up the political agenda, and in terms of creating 'visions' for fringe areas at the local level. There is a consensus that policy could do more to ease the tensions, and promote interactions, between land uses at the fringe. Clear leadership has a role in building consensus around the need to manage the fringe more effectively, perhaps through wider application of some of the initiatives explored in Part Two. Local authorities will have an important role in steering partnerships and building this consensus. That said, 'community leadership' is a useful concept for managing the fringe: this seeks to empower people and builds on the idea that the role of local authorities is not simply to commission, deliver and manage services, but also to govern local communities (Leach and Percy-Smith, 2001: 78). 'Governance' is not merely about representation, but about developing local voices: local authorities remain important in terms of leadership because they retain a democratic mandate. But this is not a mandate to 'go it alone': rather, authorities should be looking to build effective partnerships with other bodies (in the public, private, voluntary and community sectors). 'Local government' might be thought of as something different from a 'local authority' which in the past has provided a paternalistic form of leadership. Local government provides negotiated governance and fits squarely with the ethos of spatial planning. Past relationships at the fringe have often been adversarial given the strong competition for space. Governance and leadership built on partnership provide a context for easing past tensions.

Partnership
Consensus building, in any context, will involve bringing people and interest groups together in some form of partnership. Consensus is achieved when partners feel that they are part of the process and that their interests are being represented. In a planning context, and where aspirations for particular land sites are concerned, consensus will be built on compromise and mutual appreciation of the objectives of different partners. Often it is about coping with dynamics: something land-use control has not sought to achieve (Albrechts, 2004: 745). Spatial planning is concerned with integrated development, with delivery 'through' the interests of selected stakeholders, with managing the complexity of change, and with engaging in negotiated governance. Partnership is a key dimension of any 'socio-spatial process' and part of 'community leadership'. Effective partnerships depend on trust between the public and private sectors (Newchurch and Co., 1999).

Integrated management
Closer integration between land uses, as a desired outcome, will depend on integrating interests in the planning process. In the fringe, a key challenge has been the integration of social and private interests. In the land-use model, a common difficulty has been the 'silo mentality' which can mean that individual stakeholders pursue their goals according to 'ownership' of resources, with little regard to the objectives of others. Weak coordination between different actors and strategies often leads to a lack of realism about how individual contributions work for or against achieving an integrated whole (Carmona *et al.*, 2003: 219–220). Separatism can occur within authorities and it is expressed as a lack of 'corporate working': planning authorities may seek to deliver affordable housing on the back of planning agreements without sufficient recourse to housing departments. But separatism

between 'external' partners can be more pronounced. Integrated management is an expression of negotiated governance and community leadership. It may mean, in practice, creating a core management team for fringe projects comprising key figures in different organizations and community representatives.

Inclusion

Inclusion has both 'empowerment' and 'goal' dimensions. The issue of access to the fringe has been raised at various points: but access is only viewed as important because the goal of planning and management is to create places for people. In the past, the fringe may have served communities (see Chapter 5) but it was not for communities. Inclusion is about shaping the fringe according to the aspirations of its community of users (Chapter 6). According to Carmona *et al.* (2003), the principle of inclusion in the development and planning process can take different forms. These encompass consultation exercises, exhibitions and public meetings, forums, 'Planning for Real' workshops, cross-party panels and working groups. Broadly speaking, what all these approaches have in common is that they provide a framework for discussion which enables communities to influence policies and the management plans for their area. They are the nuts and bolts of negotiated governance. In 'Planning for Real' for example, large-scale 3D models are utilized to encourage community members to provide suggestions, identify and work towards design solutions. Inclusion is not merely a process by which planning is opened up; it denotes a mindset whereby all manner of interest groups work towards an open process of longer-term agenda setting that reflects a shift from the conventional machinery of government to a process of governance.

policy intervention and planning in the fringe. For this reason, the programme is given special attention in the remainder of this chapter.

Planning reform and the rural-urban fringe

Fringe landscapes are complex: too much planning policy and practice in the past has attempted to compartmentalize land uses, creating zones of activity, and failing to promote the beneficial interactions that present themselves at the edges of cities. At the same time, planning has also failed to be forward thinking, or to pursue proactive development agendas: rather, it has sat back and responded, often in an ad hoc way, to development pressures (Carmona *et al.*, 2003). The reforms instituted through the Planning and Compulsory Purchase Act 2004 seek, in the rhetoric of the government, to instigate a 'step change' in planning practice: a move towards more positive and responsive planning that reflects community concerns, works more effectively with business and, through better urban design, is concerned with place-making. These objectives seem to bode well for the fringe, as they would appear to promote multi-agency working (accommodating competing interests) while also prioritizing the place-making agenda. Indeed, Carmona and Gallent (2004) have suggested that reforms address generic weaknesses – relevant to the fringe – in the planning service in a variety of ways (see Table 9.2).

Of particular relevance to the fringe are the participatory approaches that are encapsulated in Statements of Community Involvement, the formation of Local Strategic Partnerships and the integrated appraisal of Local Development Frameworks (LDFs). These would appear to provide opportunities for greater multi-agency working at the fringe, the development of a more integrated policy agenda, and enhanced community involvement (i.e. of communities abutting fringe areas). These all edge the planning agenda towards greater integration of landscape functions, and perhaps in the direction of a more 'multifunctional' and multifunctioning fringe. According to the government, the reform agenda is also about reskilling planning to create a 'confident and dynamic profession' (ODPM, 2001: 59), and about a proactive planning service that uses LDFs to steer development and uses growth to define a positive vision. In other words, leadership in driving a vision is critical. As well as nurturing greater involvement and improving leadership, reform is also concerned with the local tools available to authorities. Master planning is being prioritized as a means of creating strategic frameworks: but as far as the detail of local planning is concerned, action plans will be available to deal with the specifics of particular contexts. These changes present opportunities for planning at the rural-urban fringe, and it is to these that attention now turns.

Partnership and integrated management

A study by the Centre for Urban and Regional Ecology (CURE) has recently concluded that 'while there is a great diversity of initiatives in the urban fringe,

Table 9.2 Past problems, planning reform and the rural-urban fringe

Planning at the fringe	Issues/Problems	Government's response
Planning lacks vision and is a reactive, negative process	➤ Planning is under-skilled ➤ Planning lacks a strategic focus ➤ Local plans, in their current form, have proven too inflexible	➤ Continuous review of core LDF policies and annual review of action plans [at the fringe]; ➤ LDF to steer 'development and use growth to define the vision for their areas' ➤ Action plans for town centres, neighbourhoods, villages [and fringes]: could take the form of area master plans or design statements ➤ Clear strategy for defining LDF objectives ➤ Promotion of master planning to improve the quality of development; master plans also to be used to ensure that concepts are translated into reality [multifunctionality?] ➤ Simplify CPO powers for land assembly [at the fringe?] ➤ Re-skill planning to create a 'confident and dynamic profession'

Planning reform and the rural-urban fringe 189

It is an exclusive process that fails to engage with communities and other stakeholders	➤ Planning is remote, hard to understand and difficult to access ➤ Communities feel detached from the planning process. It fails to engage communities, and others, because of protracted processes that people give up on, because decisions are sometimes made without proper stakeholder engagement, and because legal procedures are socially exclusive	➤ LDFs to have a stronger element of community participation, including statements of community participation; compliance to a statement will be a material consideration on larger developments ➤ Greater input into action plans [at the fringe], around which involvement will be focused ➤ LPAs to work with strategic local partnerships [at the fringe] to establish mechanisms for involvement ➤ Offer community groups advice on planning issues and further support the planning aid network ➤ Create opportunities for communities to engage prior to applications [or strategic decision making] in line with statements of community involvement ➤ Keep communities informed using e-planning ➤ Planning communities discussing applications to be held in public and best value inspectors to view closed meetings of this type as contrary to principles of best value
There is insufficient integration between agencies and policy areas	➤ Different levels of planning are inconsistent with one another ➤ There are inconsistencies in national guidance ➤ Policy areas remain divided by silo mentalities	➤ LDFs to contain statement on how they take account of other policy areas ➤ Possibility of authorities preparing joint LDFs [at the fringe] ➤ Continuous updating of LDFs to ensure they are consistent with national and regional policy ➤ Abolition of Structure Plans ➤ Integrated appraisal of LDFs covering economic, environmental and social impacts ➤ New Regional Spatial Strategies (RSS) with statutory status: LDFs to be consistent with RSSs – a requirement set out in PPS 11 ➤ Refocus national guidance (in the form of planning statements) making it less prescriptive and more able to adapt to local circumstance [including fringe areas]

Adapted from Carmona and Gallent, 2004: 111

these are for the most part fragmented and uncoordinated' (CURE, 2002: 113). In 1999, the Countryside Agency called on the government to encourage local authorities to prepare what it termed 'greenscape programmes' for all major towns and cities that would 'bring together public, private and voluntary partnership in a long term strategy for the integrated management of countryside in and around their area' (ibid.: 1). Three years later, CURE proposed that better policy coordination and integrated management might be more easily achieved if a 'Countryside around Towns Planning Framework' were to be established (CURE, 2002: 116–124). Indeed, policy coordination – nationally and perhaps through the regional planning framework – would provide the necessary context for local coordination and action. Land Use Consultants have argued that ongoing reforms of the planning system are bringing 'opportunities for more strategic, spatial, positive and locally responsive planning' (Land Use Consultants, 2003). Indeed – and despite the abolition of structure planning – there is now far greater concern at the regional level for the development of sub-regional strategies. Many Regional Assemblies are currently in the early stages of drafting Regional Spatial Strategies (RSSs), and are looking closely at criteria-based policies for rural areas and, in some instances, for the countryside around towns (see Bianconi et al., 2006). These may eventually provide a more strategic context for spatial planning – taken forward through Local Development Frameworks – at the fringe.

There are a variety of frameworks in which multi-agency working can take place. Local Strategic Partnerships (LSPs) may come together to take forward Action Plans for fringe sites. In the context of planning reform, these partnerships:

> provide the vehicle for the implementation of Community Strategies as well as the process of 'creating sustainable communities'. Local partners working through a LSP are expected to act strategically to deliver decisions and actions which join up partners' activities across a range of issues, enabling each of them to meet their own target and goals and tackle cross-cutting issues more effectively.
> (Landscape Design Associates, 2004b: 12)

LSPs may become a key delivery mechanism at the rural-urban fringe, providing a coordinating framework and bringing together economic, social and environmental objectives, tying together the emergent LDFs with specific local initiatives. Hertfordshire County Council together with several districts and adjacent unitary authorities are currently in the process of establishing Strategic Partnerships as vehicles for taking forward the county's suite of Community Strategies. Hertsmere Borough – which covers a significant part of the Watling Chase Community Forest – has developed an LSP that aims to integrate with its established Community Forest Partnership. The new Partnership will oversee the Borough's Greenways Strategy (see Chapter 7) and will start working towards access and biodiversity enhancement. There is a hope that the LSP will broaden the scope of the established Forest Vision and deal with a wider array of fringe concerns (Landscape Design Associates,

2004b: 13–14). The LSPs may well become a critical framework in which to develop the linkages needed to support the 'integrated thinking' that might eventually deliver more 'integrated' outcomes on the ground. Indeed:

> LSPs are key to improving social cohesion, the relationship between different communities in an area and their relationship with statutory authorities. They also strengthen connections with, and between public sector agencies, local government, the voluntary sectors, businesses and local residents. Overall, LSPs ensure public services work better and are delivered in a way that really meets the needs of local people, and that economic, social and physical regeneration is sustained, in both deprived and prosperous areas.
>
> (ibid.: 13)

Alternatively, special arrangements might be put in place along the lines of those used for the promotion of Community Forests. The government's Rural White Paper (HM Government, 2000) called for the Community Forest approach – which, as we saw in Chapter 7, aims to regenerate the edges of built-up areas through local participation in environmental initiatives – to be adopted more widely and for greater consideration to be given to how it might be used to assist with the implementation of other regeneration, forestry and community based initiatives. In response, the Countryside Agency set up the REACT (Regeneration through Environmental Action) programme in 2002, which attempts to incorporate environmental regeneration alongside health, sports and educational objectives. It does this by integrating with the work of Area Based Initiatives including Health Action Zones, Sports Action Zones and Education Achievement Zones (Entec UK, 2003a: 15). Hence, the REACT programme tries to demonstrate how 'multi-objective' or multifunctional partnerships can be created for the benefit of fringe and non-fringe areas alike.

How these partnerships operate and the way different interests are represented is a critical issue. Some working principles have been set out by Entec UK, following their analysis of the REACT programme. These include the need to be certain of the commitment of the project partners; setting a precise project focus with clear objectives, even if precise targets are not set against these; identifying and demonstrating where added value can be achieved either through complementing existing projects or contributing to the achievement of wider local authority performance targets such as through Best Value Performance Indicators or Public Service Agreement targets (these are set locally, so national guidance on where or how these might be met is inappropriate); the creative use of resources to undertake work such as aspects of community engagement not previously tackled; and, finally, the use of a wide range of media to publicize the project to fellow professionals, local communities and beyond (Entec UK, 2003a: 75–76). Vincent and Gorbing Ltd (2004: 32) have argued that the success of any partnership – though their particular concern is partnerships established to promote and develop community green spaces – will depend on 'organisational capacity'.

Past experience suggests that much has to 'go right' to make partnerships function efficiently, and great importance is attached to both community involvement (giving the process greater legitimacy) and leadership (providing the 'impetus for success').

Community involvement

The fringe is generally a 'landscape without community' (Chapter 6), but has a community of users and there are, of course, residential communities in areas abutting the urban edge which have a sense of ownership in what happens in the fringe beyond their homes (see Shoard, 2002: 127). Community involvement can be viewed as part of a broader 'multi-agency' or 'inclusive' planning approach, but the peculiarities of engaging with non-professional stakeholders perhaps deserve special attention. The practice and processes of community involvement – in relation to derelict land reclamation adjacent to communities and at the edges of towns – have recently been examined by TEP and Vision 21 (2004) who designed their own framework for engagement, and concluded that 'the sustainability of improvements [through land reclamation] is strongly linked to the level of public participation in the reclamation process' (TEP and Vision 21, 2004: Para. 4.18). The study further concluded that levels of community involvement tended to be greater in those cases where reclamation objectives were relatively complex and wide-ranging; that social, educational, health and economic benefits are particularly influenced by community involvement, with greater positive impact derived from higher levels of participation; that community involvement that was stakeholder-led (or led at a higher level) necessitated the existence of a formal residents group, which has further advantages for development of community cohesion; and finally, that benefits derived by securing community involvement in the reclamation process were financial as well as functional, with cost efficiencies secured in respect of site management (ibid.: Para. 5).

These conclusions appear highly relevant to planning in the fringe, for a number of reasons. First, the complexity and breadth of issues at stake in the fringe may mean that more people are interested in having some input, and should be given the opportunity to do so. Secondly, targets for social, educational, health and economic development (particularly endogenous enterprise) should be driven by local needs and aspirations, and therefore should reflect the desires of local communities. Thirdly, community involvement is likely to be initiated and orchestrated by a key stakeholder – perhaps the local authority – and this can be achieved more easily where formal residents groups are established (see also, Carmona et al., 2001, 2003). Resident groups also foster greater ownership in processes and outcomes. And finally, many funding streams can be brought on line only where reasonable levels of community involvement can be demonstrated: it is also the case that partnerships, in general, will 'significantly enhance ability to access funding' (RSKENSR, 2004: 2.1) as experience in the Greening for Growth initiative demonstrates.

Further work by TEP and Vision 21 has suggested that the processes and procedures of community involvement might be incorporated into a 'decision support framework' in which the aims of involvement were specified clearly in any partnership agreement; this might be formalized, perhaps in the same way as Statements of Community Involvement will be integrated into LDFs (it is also the case that much involvement in the planning and management of the rural-urban fringe will, in fact, occur within the scope of these statements). Involvement might also be formalized within the programming of works, and the commitment to involvement should be long-term, including monitoring of how the community subsequently uses a site. TES and Vision 21 conclude that much can be achieved in the context of Local Strategic Partnerships: community involvement, though deserving special attention, needs to be seen as part of a multi-stakeholder delivery mechanism at the rural-urban fringe.

Leadership

We noted earlier that 'leadership is important both in terms of pushing the fringe up the political agenda, and in terms of creating "visions" for fringe areas at the local level'. A national leadership role is already being played by the Countryside Agency – which will formally become part of Natural England in 2007 – though its emphasis on a 'greening' agenda in the fringe does not have universal support. In terms of local leadership, the democratic mandate of local authorities is likely to mean that they will play a key part in pulling together local partnerships who might, in an ideal world, collectively lead on fringe planning. How this process might work, and the problems it might encounter, are in part revealed by recent experience in the Community Forests programme. Much of what the Countryside Agency wants to achieve generally for the rural-urban fringe – the Agency has published a glossy leaflet setting out a 'vision' which would see the fringe become tidier, greener and more accessible (Countryside Agency, 2005) – already provides the rationale for the Community Forests programme, which is 'widely viewed as a success, [remaining] at the forefront of delivery within the rural-urban fringe by: "spearheading the approach to a multi-functional countryside in and around towns and cities; delivering in an integrated way across environmental, social and economic agendas; and operating in areas of greatest public need and opportunity"' (Countryside Agency, 2003: 1).

A more comprehensive evaluation of the Community Forests programme is currently underway, which may of course reveal that the programme has faced more problems and barriers than the Agency cares to admit. The evaluations that have been completed to date reveal some problems in the management of the programme, and provide clues as to the form of leadership necessary to take forward an alternative model for the fringe. First, individual Forest initiatives are driven by partnerships comprising the Agency, the Forestry Commission and local authorities within the Forest area. Funding to the individual Forest programme is split equally between the Agency and the incumbent authorities. However, a preliminary evaluation in 2003 revealed that effective implementation is largely dependent on the willingness of local authorities to engage with the programme and provide clear leadership. Authorities need to buy into 'the vision', and then carry it forward. This leads into a second problem. It has often proven difficult to bridge the gap between the vision for the Forest within the Forest Plan and local plan policy and decision making. For example, there tends to be uneasiness towards the use of planning gain (Section 106 Agreements) as a mechanism for securing woodland planting. This suggests that the ideals of the vision have not been fully accepted on the ground; the Agency is reliant on local authorities to drive the programme, but some authorities remain sceptical about the choice and availability of tools for meeting the agreed objectives. It has been argued, for this reason, that development plans and development planning are not providing the strong leadership that the programme needs (see, for example, Oxford Brookes University, 2002: Para 3.28).

Part of the problem might be that the Agency was too optimistic over what might be achieved on privately owned land (a problem highlighted in Chapter 7). Eighty per cent of land covered by designated Community Forests is privately owned, and the majority of tree planting has occurred on the remaining 20 per cent in public ownership. Something must change in the structure of the programme or in the wider planning context (more likely the latter) in order to convince private interests of the benefits of signing up to any alternative 'vision' for the rural-urban fringe, which may not be wholly development-based. The experience of the Community Forests projects suggests that private interests have been generally apathetic, in part because they see little to be gained by greater involvement, but also because they have been poorly represented in local partnerships. Greater private sector involvement is pivotal to the long-term success of both the Community Forests programme, and any wider programme of change for the fringe. Indeed, any comprehensive model or vision for change in the rural-urban fringe will, by virtue of their proprietary interest in the land, need to bring private stakeholders on board at the earliest opportunity. In Chapter 7, we examined very briefly the issue of corporate social responsibility (CSR). Arguably, this provides a mechanism for increasing private sector support for socio-environmental projects, and may mean that utilities and large companies become more involved in non-commercial development in the fringe. However, fragmented land ownership in many areas and a proliferation of relatively small landholdings may limit the impact of CSR. Smaller companies may see less benefit in investing in projects designed to enhance their community image if image is not viewed as particularly important for sustained business success. Hence, a question mark remains over how small landowners – feverishly hanging on to land in the hope of future development profit – can be encouraged to participate in non-commercial activities.

Penn Associates (2003) have suggested that Compulsory Purchase Order (CPO) powers should be used to purchase land in Community Forest areas, to plant it and then to subsequently maintain it. However, the use of CPO powers to promote a wider programme of change in the rural-urban fringe is clearly not a viable option, so it is essential that public and private partners work together more effectively. But, again, it remains unclear how this might be achieved: Penn Associates notes that the Community Forest programmes should seek to engage with a wider array of partners (including regional agencies, ibid.: 40) representing a range of different interests – including private interest – but no-one broaches the issue of enthusing reluctant landowners. As we noted above, a 'vision' for what the fringe might become and its future function has been published by the Countryside Agency (with the Groundwork Trust) (2005). This is very much an extension of the Community Forest vision, established in 1989, and is largely concerned with greening the fringe, with promoting community access and 'sociocultural' experiences. It is not overly concerned with the context for planning in the fringe (a context that we explored in earlier chapters); with the new economy; with formal leisure experiences; with the role of the fringe in retail; or with its general function in providing the 'urban dowry' (though how some of these functions might be greened is touched upon). In the best tradition of 'visioning' documents, the Agency and Groundwork Trust have provided something entirely aspirational. Perhaps other bodies will issue alternative visions. It might be interesting to see what the Confederation of British Industry (CBI), the Country Landowners and Business Association, the Royal Town Planning Institute (RTPI), the Automobile Association (AA), or Friends of the Earth (FoE) came up with. Inevitably, each vision would reflect a particular agenda but would pay lip-service to other areas of concern. Hence the AA might talk about future transport demand and the need for 'environmentally sensitive' road enhancement, perhaps with integrated hedgehog crossings. The Friends of the Earth would probably focus on the richness of the fringe's habitats but add that Britain needs a 'living countryside' where people have access to jobs and decent services. The CBI might well speculate on the growth of the new economy in the fringe, helping Britain retain a competitive edge in an enlarged Europe, but noting the ecological responsibilities that come with the locational advantages of the fringe. The vision put forward by the Countryside Agency and Groundwork is no different. It reflects a particular agenda – tree planting, greening, access, school children learning about nature – but talks, where necessary, about development and economic prosperity. Visions are often as partisan as the processes creating the landscapes that they seek to change. National leadership is probably not about creating some generic vision for the rural-urban fringe: rather, it is about handing local bodies the opportunities needed, to work out for themselves, what potential their local fringes have and how this might be realized. Hence talking about leadership, partnership and inclusion, though appearing rather general, is more useful than the development of a fanciful vision that few people are likely to sign up to. Leadership is of course a matter of providing 'vision', but not necessarily a prescriptive vision: it has to be grounded in realism and achievability, and at a national level it should really be concerned with giving local agencies the tools to address local challenges.

Clearly, many tools are already available. We have already mentioned Local Strategic Partnerships, Compulsory Purchase Orders, Local Development Frameworks, the possibility of planning within an enhanced green belt or other softer local designations (Chapter 8), the model of partnership provided by REACT or the Corporate Social Responsibility agenda, or looser planning frameworks for the fringe that might encourage the development of some shared agenda. One mechanism that we have not touched on so far is the adopted 'Action Plans' emerging from planning reform. These could provide a clearer framework for intervention in the fringe without having to buy into any generic vision. Action Plans (plus other forms of designation) and other opportunities are now considered in greater detail.

Planning tools at the edge

The designation of rural-urban fringe areas for special treatment may become a viable option through planning reform. Bickmore Associates (2003) point out that:

> Many writers, researchers and planners have been aware of the problems of the 'urban fringe' for some years and many ideas have been put forward for their solution. . . . Nearly all of these ideas depend on a special designation being established for the 'urban fringe', special powers acquired for dealing with such areas, and special treatment being given to it. However to date none of these propositions have been accepted into the land use planning system. Government contends that the existing provisions within the planning system are adequate and extra designations would be of little or no benefit.
>
> (ibid.: Para. 7.30)

It might be easier for some of the suggestions to be incorporated into a new 'sociospatial' approach to planning: the land-use model tended to reject extension into what it considered to be 'non-land-use' issues, which presented quite a barrier to innovation. For instance, any concern for agriculture was rejected by the Town and Country Planning Act of 1947: 'cultivation' is not one of the land uses identified in the Act's definition of development (see Chapter 8), or indeed one of the land uses described in the most recent Use Classes Order (2005). But agriculture – despite its fragmentation – remains an important activity in the fringe, and Bickmore Associates (2003) argue that where farming is becoming marginalized, and less viable, the planning system should be able to give it proper support, or enable movement into other economic activities (ibid.: Para. 7.13). A key problem is that planning is able to influence what goes on around farming – which may have a profound impact on its viability – but it fails to understand the changing economic context for farming, shifting practice, or alternative management options. The same is true in relation to many other functions in the fringe, especially ecology (Chapter 7). Public-sector-led planning is a model drawing on a limited skills base, relying too much on the capacity of planning officers to appreciate the needs of all sectors: to be 'Jacks of all trades but masters of none'. The first step in designing (and implementing) appropriate planning tools at the fringe is to understand the context. From planning's point of view, this may mean relinquishing or sharing management responsibility: in relation to agriculture, Bickmore Associates (2003) call for an approach that:

- Acknowledges the special challenges and opportunities that come together at the fringe;
- Understands the forces that are giving rise to shifts in agriculture (as a major fringe land-use) and the move to non-agricultural activity or to non-food crops;
- Understands the changes in the functioning of the rural-urban fringe and the way this functioning is reflected in the landscape, and how planning can influence outcomes;
- Understands how changes are viewed by society at large, and accepts the range of concerns that people feel, from visual, aesthetic, environmental and conservation standpoints;
- Accepts that the regulatory regime – i.e. the planning process – is to some extent a mirror to these standpoints. It is the child of the post-war period when the prevailing view was that the countryside should be 'timeless'. Planning therefore fails to promote or manage change effectively, perhaps making it unsuited to facing the challenges presented at the fringe;
- Recognizes declining agricultural production in some areas, especially in fringe areas, where other uses are sometimes vying to step into farming's shoes, which means that planning must be able to handle and plan for rapid change effectively;
- Recognizes that planning needs to acknowledge the interplay between different uses, and especially between farming and non-farming uses. It needs to be implemented flexibly, and with discretion (a view that runs contrary to the objectives of green belt policy, but which is very much in line with the thinking discussed in Chapter 8). It also needs to be combined with investment to promote and ease the path of change (see BK Consultants, 2003); and
- Accepts that any planned strategy needs to embrace farming concerns, acknowledging the key role that farming will continue to play in the landscape of the fringe.

These points suggest a need for some up-front investment in understanding what is special about the fringe. Like all landscapes, it is 'made', but generic political, economic, social and environmental processes have tended to act upon the fringe in a particular way (Chapter 2). The interplay between forces and functions is different – as Bickmore Associates acknowledge – and this needs to be understood by those wanting to intervene in the 'natural' processes making the landscape. It is not so much about doing something 'special' at the fringe, as ensuring that where policy interventions are implemented, these are grounded in understanding. One way forward might be to establish 'action plans' for areas of fringe, with 'step one' of any plan being an intention to appraise the local context, and to understand the interplay between the fringe's key functions in a particular setting. Entec UK (2003b) have proposed that 'urban fringe action plans' (UFAPs) should be implemented in the context of Local Development Frameworks. A study looking into this issue in 2003 was less concerned with the principles (apart from achieving 'multifunctionality') that might underpin any local framework, and more concerned with practicalities. Entec concluded that UFAPs would fit comfortably in the new planning system and could be used to articulate a clear and practical vision; would be coordinated by partnership working; would draw on stakeholder and local community involvement; would adopt a 'multi-functional philosophy'; would nest within the 'strategic planning framework'; would include a 'delivery plan'; would be coordinated across boundaries through a partnership agreement; and would contain both land-use and land management proposals (adapted from Entec UK, 2003b: 18–19). This is a model that covers many of the issues and principles highlighted earlier in this chapter. Nesting action plans in a strategic framework (ensuring compliance with Regional Spatial Strategies) suggests that they might cover fairly sizeable areas, and might well be 'cross-boundary' in many areas. The South East Plan (one of the first Regional Strategies to emerge in draft form; South East England Regional Assembly, 2005) already identifies areas of 'London's fringe' though the definition of fringe is rather broader than that adopted in this book. It covers areas of suburban sprawl extending beyond the boundaries of Greater London. But the designation of London's fringe as a 'Growth Area' in the South East Plan suggests that if Regional Planning becomes involved in identifying fringe areas – working with those putting together action plans – then we might well expect to see areas of fringe figure more prominently as areas with 'special needs' within future planning policy, perhaps at all levels.

Admittedly, the objectives – centring on partnership, strategic thinking and a widening of the role of planning – of Entec UK's proposed UFAPs are so broad that they could provide a framework for any action plan designed for any type of area. Within the current nomenclature of spatial planning, they 'push the right buttons'. But in light of the context for planning in the fringe, these objectives appear especially important: and, at the very least, the arrival of fringe action plans would mark an end to the neglect that some fringe areas have experienced in recent times. There are also more specific proposals in the Entec study: there is a suggestion, for example, that community strategies (part of the Local Development Framework documents) should concern themselves with clusters of communities located in the fringe (ibid.: 24). This might mean looking at the needs of the 'hub estates' discussed in Chapter 6, or paying particular attention to the pressures facing communities 'washed over' by green belts (Elson, 1986: 134). As we noted earlier, the Countryside Agency has already developed its own 'vision' for the rural-urban fringe (though it has now replaced the term 'rural-urban fringe' with 'the countryside in and around towns'), but has shied away from suggesting exactly how this vision might be realized. Entec have called on the Agency to make action plans a central delivery mechanism, adding that 'sub-regional planning' – orchestrated at the regional level – might become the key arena for the implementation of these plans. Work by Oxford Brookes University (2002) adds support to the idea that regional planning has an important role to play in the fringe. 'Regional development strategies' produced by the Regional Development Agencies in England have, for example, been instrumental in directing EU Objective 1 funding to 'strategic greening sites'. These strategies might provide a useful sub-regional framework for area action plans. The wisdom of operating at this scale is, according to Entec, confirmed by past successes in implementing 'river valley projects' and Community Forest initiatives (Entec UK, 2003b: 40). These have not been constricted by boundary issues, and neither should fringe initiatives designed to deliver greater connectivity to urban areas, as well as a suite of social, physical and economic enhancements (ibid.: 4) that require neighbouring authorities to work together. There is little in the Entec study to disagree with. UFAPs would focus attention on the fringe; they would channel resources, where available and necessary; they would provide a framework for partnership, integrated management and community involvement; and with the move away from old-style development plans, they present an opportunity to embed fringe concerns within the planning framework. They also appear to sit comfortably with the government's Communities Plan (ODPM, 2003a) with its embedded 'liveability agenda'. In the Thames Gateway, for instance, this agenda is expressed in 'Greening the Gateway' (ODPM, 2004b) which aims to integrate local – including fringe – landscape concerns with the goal of achieving physical growth. The liveability agenda could certainly be taken forward in urban fringe action plans tied to particular communities or parts of the Gateway development. Perhaps the central role of action plans would be to send out a clear signal: that the fringe is no longer out of sight and out of mind.

There are many other planning issues that are potentially important at the fringe that have not received attention in this chapter. The future of 'planning obligations', for example, is an issue picked up by the ODPM in its consultation on 'Contributing to Sustainable Communities' (2003c). The move to a 'twin track' approach where local authorities set a local 'planning charge' (tariff) linked to residential and commercial development would remove the requirement of 'planning gain' being linked directly to a development site. This might free up more planning gain income for fringe projects, assuming the fringe can out-compete education, health, and housing for a slice of local development derived revenue. Another key planning issue at the fringe is the future of the green belt. Landscape Design Associates (2004b: 6) have noted that 'potential reforms to the green belt policy in respect of how this designated land is controlled and managed at the interface with settlements (i.e. the urban rural fringe) and used by local communities' are likely to have a profound impact on Community Forest, and broader fringe strategies. In Chapter 8 we suggested that containment at the fringe has not always resulted in the most advantageous economic, social and environmental outcomes: reform of green belt policy is long overdue.

With reform of the planning system in England comes an opportunity to do more at the rural-urban fringe. But focusing attention on the fringe, and planning with greater care, is very different from the notion that fringes should be transformed into something entirely different. It is, arguably, eminently sensible to cluster development rather than allowing it to litter the landscape and consume open space; it is also logical to bring different functions together where they are compatible and might share mutual benefit. Improving access, lightening the footprint of new development, and maintaining biodiversity: these are all laudable aims. But it should also be remembered that a great many people find value in the relative wildness of the fringe, though they often express support for intervention that removes the threat of environmentally malign development in the form of roads truncating important habitats, or the arrival of yet more red-brick boxes spilling over into the fringe. There is consensus that planning has a role in the rural-urban fringe, but this ranges from Shoard's view that planners should be 'handmaidens' helping along 'a universe' which they do not seek to control (Shoard, 2002: 140) to those aspiring to create a well signposted parkland dotted with exemplar projects for sustainable development, all neatly master-planned and urban designed.

Conclusions

The claim that land-use planning has sought to integrate and harmonize various space-using functions can be challenged at the rural-urban fringe. The focus of this model has, to a significant extent, remained on control and regulation. At the fringe, this focus has resulted in a specific range of urban uses being pushed out of cities and brought into conflict and competition with those uses more usually identified with the countryside. The land-use patterning of the rural-urban fringe – noted in Chapter 1 – is one of disparate uses rarely harmonized with each other or their landscape context. Spatial planning provides an opportunity to challenge the inert and passive nature of fringe planning. However, spatial planning is not new: indeed, strategic spatial thinking has been a goal of structure planning since the 1960s and integral to new-style regional planning since the early 1990s. It is not so much the concept of 'spatial strategy' that is useful for the rural-urban fringe, but rather what this concept means in practice, and how it might be taken forward. Everything that planning at the edge has hitherto lacked – long-term vision, integrated development, the management of rapid and complex change, and negotiated delivery – is neatly parcelled into spatial planning. In a sense, it merely demonstrates what planning has failed to do and what it might do on the back of greater political will and commitment, adequate resourcing, and more effective public–private collaboration.

But the issue raised at the close of the last section – the required extent and objectives of planning at the edge – is, of course, the critical sticking point. As we noted, accepting that planning should pay greater attention to the fringe and understand its dynamics, is very different from the idea that it should seek fundamental change. This is perhaps the critical battle-line within England's rural-urban fringe: on the one side, there are those who wish to retain the fringe's 'unplanned character' and, on the other, those who 'have a vision'. Thankfully, a national strategy for 'sanitizing' the fringe is unlikely to emerge: rather, local groups will be left to their own devices, to work out for themselves what needs to be done, and to develop solutions that reflect local needs and aspirations. Making the fringe a higher priority locally, developing skills, securing funding if needed, finding lasting solutions, and working with landowners: clearly there is much that will still need to be worked out. But surely providing a framework for local action is the best solution. The diversity of the rural-urban fringe will be retained and the risk of it being theme-parked will recede.

10

Conclusions

Introduction

The 'countryside around towns weaves in and out of the built up area and is often characterised by new development, derelict and brownfield sites, retail and industrial parks, land fill sites and reservoirs' (Countryside Agency, 2005: 4). In this book we have tried to add detail to the general image of the rural-urban fringe that can be gleaned from any train journey, or any trip between airport and hotel, in almost any city in the world. This anonymous, universal landscape has a dynamism that is often hidden from the casual observer. It is a landscape of intense land-use competition; with a historical and cultural heritage gained from past service to the built-up area it envelops; with a biological richness that is only now being recognized; offering a diversity of sociocultural opportunity; and home to a new service economy comprising basic functions ranging from distribution and storage to high-tech research and development. Many activities and functions find a home in the rural-urban fringe.

But the future of the fringe is now up for grabs, and a conflict is brewing between those who celebrate its anarchy, and those who view its unconventionalism with disdain. In England, a vision of the future fringe has been put to the government which seeks to promote fundamental change: a fringe that is no longer wild and unkempt, but becomes more like the countryside beyond, yet shares the same concern for good urban design as the city within. As we have noted at various points, and emphasized in Chapter 9, having the principles in place with which to frame local action is very different from having an 'over-designed' objective in mind. In our view, some agencies have gone too far down the latter path while, on the other hand, certain independent commentators continue to promote a view of the fringe that plays only to 'middle-class' intellectual desires. But having principles in mind to guide local action is essential. There would be little point in offering this contribution to planning at the rural-urban fringe, if we were not of the view that policy intervention – however extensive or subtle – does not need to be framed in an understanding of context. The review of context provided in this book suggests a fringe of apparent tension, but potential compromise; a fringe where a multitude of groups have actual or potential interest in future development; a fragmented fringe in terms of landholding, political influence and private aspiration; and a fringe that society at large fails to understand and therefore undervalues. Even if we remove the need to develop some 'generic vision', we are left with the need to guide future action. In the last chapter, we argued that fairly simple guiding principles will suffice:

- Clear leadership: in drawing attention to the fringe nationally; in developing a strategic framework regionally; and in delivering outcomes locally;
- Partnership: multi-agency working, possibly through 'local strategic partnerships' or through other special arrangements; drawing up clear partnership agreements, setting objectives, milestones and responsibilities; we also noted that private sector involvement in these partnerships, and in core management, is vital;
- Integrated management: bringing together different interests, coordinating activity, perhaps through a core management group; and developing a stronger sense of ownership in the fringe;
- Inclusion: involving and including community and resident groups in innovative ways; monitoring the success of involvement during the programme of delivery and subsequent use of sites.

These are clearly so broad that they could underpin almost any outcome. They fit with the view that planning could simply maintain and promote the best of the status quo, but avoid unnecessary conflicts and halt inappropriate development (Shoard, 2002: 140). Or, bodies such as England's Countryside Agency could use the principles to guide a programme of more radical sanitization. In 2002, the Centre for Urban and Regional Ecology listed a number of fairly simple objectives for policy intervention at the rural-urban fringe, suggesting that incumbent authorities should ensure that these landscapes:

- Offer diverse opportunities for improving quality of life for all;
- Embrace the idea of and deliver a strategically planned green space network;
- Display land management that maximizes benefits;
- Incorporate appropriate new development;
- Demonstrate multifunctionality (that is the plural and simultaneous use for a variety of purposes, combining a variety of qualities);
- Encourage alternative, perhaps novel, land uses;
- Have urban edges structured for connectivity;
- Are attractive, accessible (to all) and economically thriving.

These are broad aspirations: diversify opportunities where necessary; embrace strategic thinking; maximize benefits; demonstrate multifunctionality; ensure connectivity; and focus on access and the economy, where these are underdeveloped, but could be enhanced. In reference to the Countryside

Agency's 'draft vision' for the fringe published in 2004, the Royal Town Planning Institute has suggested that 'simplicity is no bad thing', but that the Agency's vision, in particular, tends 'to paint a picture of an ideal world rather than a hard-edged, practical way forward' (RTPI, 2004: 2). This is a key problem with all visions, but especially those that attempt to cut through local complexity and arrive at something very general. The solution is of course to either come up with 'pick and mix' answers – if you have situation 'A', choose option 'B'; if you have situation 'C plus D', then option 'F' is the best way forward – or alternatively, to avoid any type of prescription. Our focus in this book has been on the context for planning at the rural-urban fringe, and it was never our intention to 'envision' or predict what the fringe might become. Having said that, it is clear from our review of the context that 'improvements' might be instigated in some fringes: this is a view shared by researchers at the Centre for Urban and Regional Ecology (CURE, 2002). Improvements can also be painted as aspirations and we turn to these briefly in the next section.

Delivering change – where necessary

Some fringes are in dire straits: they are no-go areas for local communities; they are contaminated, fraught with danger, noisy and noxious. Some are performing reasonably well on an economic front, but could do better. Others have ecological or sociocultural potential that is already being realized through local projects. Some, quite simply, do not need any form of policy steerage. There is no single 'challenge' at the fringe. These are diverse landscapes; they defy simple tagging; they are unique, though they may not always appear so. This means that policy needs to deal in aspirations, which many fringes – through the work of local partnerships – may need to achieve, though some may already have done so. In Chapter 9, we talked about the integration of different land uses and functions; this has been a recurring theme throughout this book and is described in the CURE study (2002) as the 'plural and simultaneous use' of land. Our review of the context for planning reveals that many such 'plural' uses are possible. In the case of Howdon Wetlands (see Chapter 7) a sewage treatment works was rescued from near-dereliction through a project that sought to broaden its function for recreation and education. A similar scheme at Van Diemen's Land near Bedford (also Chapter 7) breathed life into a previously sterile fringe by combining a new ecological function – the creation of a wet-woodland – with an old drainage function. Something that people previously ignored, suddenly became something of interest: a resource for nearby communities and passers-by. The same is true at Colliers Moss (see Chapter 1): from economic redundancy and decay sprang a new, but very different, economic function together with opportunities for recreation and study. A particular strength in this instance is that the site does not limit recreation to an appreciation of nature: motor-cross is also welcomed, reflecting the real interests of nearby residents. This new plurality brings a potentially lasting solution to the former problems these fringes faced. Previously, the loss of the land's one and only function might subsequently mean abandonment and dereliction. Integration means that uses stand together, and gain mutual support: recreation works because there's an ecological function to retain the interest of visitors; economic uses support the historic heritage by reusing old buildings; and these economic uses are supported because businesses are encouraged to invest in a more dynamic setting, attractive to clients and employees alike. And plurality means that the fringe has more than one string to its bow: it accommodates the coming and going of different uses; it has strength in diversity. Antrop (1998) along with Bastian and Roder (1998) make exactly these points: an integrated approach to policy and partnership at the rural-urban fringe, leading to integrated and plural outcomes, is more likely to present a lasting and hence a 'sustainable' solution to the problems faced by those fringes that have endured economic and environmental decay in recent years.

However, there are some very basic barriers to achieving 'multifunctionality', or even just a simple mixing of uses. For example, a great many developers who operate in fringe areas are unfamiliar with the concept of mixing uses in a creative way: they are used to doing their own thing, in a far more fragmented manner. They have structures of land acquisition, development procurement and design which are geared towards providing single function uses on sites. In addition, the established zoning approach to land allocations – part of the prevailing land-use planning system – accentuates this problem (see Chapters 8 and 9). For example, employment, residential and education land allocations are often zoned on separate sites with few connections between them. Consequently the places created as a result do not always 'function' well. Of course, we expressed considerable hope in Chapter 9 that this situation will change: spatial planning may provide a context for integration, though it may take some time for planners and developers to let go of established working practices.

Encouraging different functions to 'interact' (as well as just share the same space) might be another aspiration of planning intervention at the fringe. As we have noted previously, many land uses at the fringe have endured an adversarial relationship with their neighbours: the local reservoir has been closed to visitors; roads have been viewed as intrusive; economic uses have detracted from landscape quality; and recreation has been barred on agricultural land. Hasler and Kjellerup (2000) make the broad assertion that policies need to 'nurture a symbiotic relationship' between functions: they need to operate within a 'multifunctional' framework, aiming to bring land uses together. However, it is not at all clear how symbiotic relationships might be achieved in practice: simply lumping uses together in the hope that some interaction might result is of course one option. Another option is to develop longer-term management strategies that aim to stimulate interaction. In the case of the Community Forest Partnerships (see Chapters 8 and 9), forums are in place that aim to involve local people in ecological projects; businesses are also encouraged to work with local communities towards 'environmental regeneration' goals. Admittedly, examples of management promoting longer-term interaction are all too rare. But in the case of the Red Rose Forest Trust (Greater Manchester), a partnership with the Co-operative Bank has resulted in the 'Co-operative Bank Community Woodland'. This project began as a concept in 1997, to celebrate the bank's 125th anniversary: with the Red Rose Forest and other partners, 'four predominantly barren areas were selected and rebuilt utilising green waste to rebuild the soil, combined with knowledge, skills and a lot of hard work to create sites that support expansive and varied woodlands' (http://www.communityforest.org.uk/partnership.html; accessed 18 June 2005). Biological development within the Woodland has been, and will continue to be, closely monitored. In this instance, a management plan has been put in place that will ensure continued interaction between, what we have termed earlier in this book, the aesthetic, the sociocultural, the ecological, and – more indirectly – the economic fringe. But some longer-term interactions appear to work well without management plans or a guiding hand. In the case of Wellesbourne Airfield, an example that we have referred to on several occasions – and other airfields across England for that matter – a range of economic and socio-economic uses come together for mutual benefit on a single site. With the end of military operations in 1964, the former RAF Wellesbourne began a new life, first as a vehicle testing ground and subsequently – from 1981 – as a commercial flying site, hosting training schools offering pilot instruction and aircraft hire. In more recent times, Wellesbourne has become home to a diversity of activities. Some land has been returned to livestock grazing; some has been sold off for housing; one of the former runways hosts a weekly market; motor-sports events and fairs are held on the site; there is also an assortment of economic uses including vehicle storage, distribution and light manufacturing. The 'interaction' between uses comes from the fact that recreational flying is only possible on the site because the owner-operator derives an income from these other uses, and is able to cross-subsidize flying activity, a visitor centre and a café. Interaction in this case is a result of the owner's desire to see flying continue, and his control over the other functions that locate on the site. At Wellesbourne, we see an assemblage of compatible functions: and the site 'works' as an integrated whole. Arguably, the movement of money is central to successful 'relationships', with more desirable 'luxury' uses being supported by more basic functions. In the case of Barn Elms in south-west London, the development of a wetland centre was possible because part of the site was used for housing and a portion of the land-value uplift was extracted through a planning obligation. But does this suggest real interaction, or simply a marriage of convenience? Funding is clearly important for many fringe projects, but the trick seems to be to convince landowners, developers and business interests that there are longer-term benefits to be gained from involvement and interaction in sociocultural and environmental projects. This is something that the corporate social responsibility (CSR) agenda seeks to achieve. As opposed to simply 'enduring' the needs of local communities and the environment, by making the basic concessions or delivering the minimum required gain, business interests are asked to consider what long-term involvement with communities and environmental projects will mean for their corporate image. If more businesses buy into the CSR agenda, then it might be possible to foster a greater range of functional interactions in the future. We return to this issue later.

At least as critical as the way that functions come together and interact is the issue of accessibility to the rural-urban fringe. The structure of any landscape is formed by the pattern of elements that make it up including roads, footpaths, cycle-ways and hedges. This structure determines how easy an area is to access, understand and navigate, determining legibility. Throughout this book we have painted a picture of the fringe as a landscape with qualities likely to appeal to a range of potential users, but either because of physical barriers, or legal impediments (see Chapter 7), it may be impossible to access the open spaces of the fringe. We have already focused on access at some length, concentrating on the way the landscape may be truncated by communications infrastructure or closed off to public access by private landowners. One further issue is the 'permeability' of the urban edge itself. Lynch (1960) argues that the legibility or 'imageability' of the city is dependent on clearly identifiable 'paths', 'edges', 'districts', 'nodes' and 'landmarks' (ibid.: 46). Of these elements 'landmarks', 'nodes' and discernable 'districts' are largely absent from the fringe and instead 'edges' – often impenetrable linear boundaries that separate areas of land from each other – are the most prominent feature. Although motorways, railway tracks and canals can be considered as 'paths', they often prevent people from moving freely and easily within the fringe and contribute to a fragmented landscape. Similarly, transport hubs are certainly 'nodes' but they may afford access only to a retail centre or a housing estate, not to the open spaces beyond. In addition, culs-de-sac (at the immediate urban edge) and scruffy, 'un-neighbourly' land uses such as waste disposal and illegal fly-tipping add to the perception of an inaccessible landscape. Peripheral movement routes (ring roads and so forth) have been built, which cut off traditional radial patterns of movement from the urban to rural; and areas between peripheral routes and the urban edge have been in-filled with development mainly accessed by car. It may well be impossible to replicate the 'image of the city' in the fringe without urbanizing or theme-parking the landscape, but it is possible to improve both legibility of, and accessibility to, the fringe in modest ways and in particular locations. This might be achieved through the following:

- Facilitating permeability at the edge of built-up areas, improving access to the fringe by avoiding dead-end or cul-de-sac street layouts that present barriers to access;
- Promoting urban forms that allow access to the fringe, bringing communities closer to the countryside around towns. For example, the use of green wedges (see Chapter 8) rather than conventional green belts might promote 'development fingers' penetrating into the fringe – a 'star'-like urban form rather than something resembling a circle: this would increase the overall length of the urban-edge perimeter, allowing more people to live adjacent to 'green fingers' penetrating into the built-up area;
- Ensuring access to fringe areas by rail and water, and through improvements to public transport, especially to transport hubs (which might become more like Lynch's 'nodes') and parkways (see Chapter 5);
- Promoting green access via cycle-ways, footpaths and 'greenways' (see

Chapter 6) of the kind developed in Stratford on Avon. There are currently 17,000 km of greenways in England and these are designed to link to 'other networks for non-motorized users – such as the National Cycle Networks, towpaths beside inland waterways, National Trails and other rights of way' (www.greenways.gov.uk; accessed 18 June 2005). In the fringe, if they linked to motorized hubs, to parkways and to leisure and retail developments, they might contribute further to permeability and accessibility;

- Addressing issues of safety and image, creating a rural-urban fringe that people value and wish to access. Part of the problem is the presentation and representation of the fringe discussed in Chapter 2; the fringe is perceived as a no-man's land and this image is confirmed through film and other media;
- Changing attitudes among ethnic and social groups who may currently under-utilize the countryside for recreation and leisure. This perhaps returns us to issues raised in Chapter 7 and earlier in this final chapter: that 'ecological recreation' tends to be the preserve of a well educated white middle class, and even with broader appeal, developed through a programme of classroom learning, other forms of recreation – weekend paintball battles, quad-biking, motorized boating, flying, or festivals – should not be subdued by those who promote a narrow view of what should and should not be permitted in the countryside. The thought of being restricted to 'quiet enjoyment' turns many people off the countryside, but who is to say that opportunities for ecological learning cannot be integrated with noisier pursuits?

These are all issues that need to be addressed in any strategy for improving access into the fringe: the expansion of green paths and public transport; linking vehicular and non-vehicular routes; making sure that bicycles can be taken easily on trains; or calming traffic at the urban edge (where possible) to encourage access by foot. These are small practical steps that might be taken to improve access and permeability and might be incorporated into the objectives of the 'action plans' for the rural-urban fringe examined in Chapter 9. But the economic and social impediments to access are also important and these tie in with the question of which 'sociocultural experiences' are thought to be legitimate in the fringe (Chapter 6). There is clearly an economic dimension to access: middle-class suburbanites with several cars to a household might be able to access the fringe fairly easily and will probably do so on a regular basis: to walk the dog; to visit a garden centre or a DIY store; to groom the family horse; or to help mend a local dry-stone wall. On the other hand, less well-off households residing in the inner city and without access to a car might find it very difficult – and fairly expensive – to take the family for a day-trip to nearby countryside. A journey by underground to the appropriate terminal overland station, followed by a half-hour rail trip, twenty minutes waiting for a bus and then another 20 minutes to reach the desired drop-off point may well dissuade all but the most determined day-trippers. They are far more likely to walk to the nearest municipal park or, if they really want to escape London or Manchester, to head to a nearby town for the day. But, of course, economics aside, the fringe is always likely to appeal most to those who live nearby and can access it quickly by whatever means: money, recreational taste, and cultural differences simply complicate matters, providing further impediments to access. On the last issue, the Black Environment Network (BEN) has suggested that 'the immediate image of Environmental . . . structures is white and therefore difficult to perceive as relevant or desirable to ethnic communities. The power for change is seen to be held in white, male, middle class hands in all areas of the countryside' (BEN, 2000: 23). This is also a problem in the rural-urban fringe. The same report adds that 'ethnic groups are in the main excluded from the official history of Britain – how do they fit into the history of British heritage, including the countryside?' Furthermore, there is a 'feeling that they are not welcomed by countryside authorities or agencies who control countryside areas' (ibid.: 23). The perception seems to be that the countryside is dominated by a white middle-class clique, probably driving around in four-wheel drives and donning green Wellington boots. Black and ethnic minority households feel particularly alienated from this image. On a more specific issue, the Greater London Authority has suggested that 'traditional forms of environmental interpretation may exclude many people from ethnic minorities' (GLA, 2002: 31). Perceived danger and personal disability also have their parts to play in limiting access to the fringe, but might be addressed through broader access improvement and educational programmes.

Achieving any of the aspirations listed above – or the goals set by CURE (2002) and others – will require a redirecting of resources and effort, at least in some areas. Of course, we are talking about a more subtle use of intervention than those agencies with grand designs for the fringe. But there will still need to be greater political support for policy interventions (aimed at improving access and so on); the development of new skills to facilitate the operation of a planning system that will rely less on public-sector control and more on partnership; funding for local projects; a commitment among different groups to achieving local goals for the fringe; and some mechanism to improve hitherto lacklustre public–private collaboration. Building political will at all levels is problematic: national government needs to be convinced of the benefits – social, environmental and economic. The Countryside Agency has championed this cause and there are some signs that the government sees the fringe as an important dimension of its 'liveability' agenda. A big problem at the local level derives from the fact that fringes are often without communities (Chapter 6) and therefore without a political voice. If voters within an electoral ward come to value their immediate fringe as a local resource, then pressure may grow to ensure that access and environmental quality are maintained. However, that ward member may be a lone voice within local government, competing for resources and attention against colleagues promoting education, crime or health agendas. The trick, quite obviously, is to 'sell' the fringe issue as integral to all these wider concerns: the fringe as a resource for local schools, as offering health benefits and as a potential arena for crime and anti-social activity if positive steps are not made to open it up and make it safer. Of course, those local councillors representing the urban edge may compete for attention and resources against a larger number of inner-urban ward members. By opening up the fringe to wider access – creating routes into it from inner areas – there is always the hope that more people will develop a stake in its future: the more people who care about the fringe, the stronger will be its political voice.

The skills needed to be creative are also in short supply within local authorities (Egan, 2004). This makes it difficult to pursue a new agenda or engage in new ways of thinking and working. CURE (2002) suggested that more 'novel' land uses should find a place in the fringe, and that plurality in land use might become the norm. This will require the development of larger networks of people with different skills, all working together towards agreed aims. In 1999, the Urban Task Force concluded that the public-sector skills base is fragmented and uneven. Four years later, a study for central government suggested that local authorities lack staff with core skills in master-planning, project management and urban design (ODPM, 2003d). This same conclusion was reached by the Egan Skills Review (Egan, 2004). These skill gaps make it difficult for local authorities to take the lead in any new policy intervention at the fringe. A lack of master-planning and design skills certainly seems to reduce the capacity of authorities to deliver on the Countryside Agency's vision of a well planned, well designed, cleaner and tidier fringe. Of course, other commentators might argue that this is no bad thing given that becoming the nation's latest urban design project is exactly what the fringe does not need: skills should be developed in order for local authorities to understand contexts, and to ensure that they are in a position to provide appropriate management. The purpose of skill development is not, necessarily, to underpin and drive fundamental change. Alongside skills, funding is also important. The Royal Town Planning Institute (2003) argues that a lack of 'mainstream funding' (ibid.: 4) makes it difficult for planning authorities to engage in 'positive' (that is, proactive) planning. Hence, they are obliged to sit back, wait for planning applications to roll in, and then simply respond to these on an ad hoc basis. This inability to invest in forward planning, and retrenchment into basic duty, are at least partly responsible for apparently sporadic, uncoordinated development in the rural-urban fringe. Central government tends to agree that funding for the planning service has been too constrained and has created a system of 'planning delivery grant' (PDG) aimed at rewarding authorities that are seen to provide a better service. A serious accusation has been levelled at PDG: it is designed to encourage speed in development control, but not angled towards delivering greater quality in planned outcomes. Clearly, a system of rewarding the innovative and creative qualities of planning – and its outcomes – might be welcomed within the fringe.

Political will, an expanded skills base and new funding sources will have a limited effect on how much can be achieved at the fringe if there is not a wider commitment to improving those fringes in need of improvement. But how can reluctant landowners be encouraged to open up land to local people? Why would they become involved in local environmental projects? What benefits might flow from a more committed relationship with the public sector? There are no easy answers. But a framework in which partnership and cooperation are encouraged would be a useful first step. Conventional containment policies in the rural-urban fringe – especially the green belt – are, paradoxically, so unbending that they generate hope in future land release. Their inflexibility during more than half a century of operation has resulted in growing physical and political pressure: physical in the sense that unprecedented growth means that some green belts are now tearing at the seams; and political in the sense that many people believe that they no longer fit with the wider ethos of planning (see Chapter 8) and its objective of delivering sustainable development. The perseverance of green belt policy beyond its sell-by date has resulted in a proliferation of agencies and web sites engaged in the brokerage of green belt land for investment. One such site recently noted that 'the market in green belt land for sale is increasing. Green belt is offered for long-term investment in the hope that it will gain planning permission to build homes' (http://www.selfbuildabc.co.uk/green-belt.htm; accessed 18 June 2005). Many major land agents now regularly offer green belt plots for sale – including FDP Savills, Strutt and Parker, Frank Knight and many others. Speculation on green belt land has been painted as an opportunity for the private individual to enjoy some of the financial rewards previously accessible only to big business: 'plot sales allow individuals to share in the development gains that previously were only on offer to the large builders and developers of land for sale' (http://www.greenbelt-news.org.uk/; accessed 18 June 2005). This view has emerged because government has granted some large releases of green belt land for 'sustainable urban extensions' and it could, indeed, be argued that this practice favours big business over the small investor. But the critical issue here is not who wins and who loses in the property game but rather, how can developments be encouraged to conform to a general set of principles for the fringe (whatever these principles might be)?

One suggestion, offered in Chapter 8, is that a replacement for the green belt – or looser designations in the fringe including wedges, gaps and buffers – might become a framework for desirable forms of development (again, whatever these might be). Thus, buildings might have to conform to particular environmental build and design standards; developments might have to achieve a range of functions; housing encroachment at the urban edge might be limited to those schemes with the highest environmental standards; and particular preference might be given to new models of living and working, including the 'cluster housing' developments described in Chapter 6. A framework for building 'commitment' may be vital, even though it might appear coercive. But commitment – and particularly a commitment to public–private partnership – may also evolve over a longer period of time as businesses slowly come to accept the importance of having a greener, friendlier corporate image. Image is not a new concept to big business: appearing modern, innovative, and reliable are clearly important to future sales. But how important is being green and caring for the environment? The fashion company Laura Ashley has a long track-record of taking social and environmental responsibility seriously: the company has a policy of relying on 'ethical supply chains'; it recycles 50 per cent of its waste products; it has a strict environmental policy and produces annual environmental reports. But profits were reported to be flat in 2002, and share prices in the company fell by 18 per cent in the following year. Product and market context are the key factors for any business. But that is not to say that companies are not beginning to take 'green' and 'social' image issues very seriously. Many businesses are signing up to the corporate social responsibility (CSR) agenda (see Chapter 9): this is a model mainly affecting larger companies and corporations at the present time, but it is likely to trickle down, in some shape or form, to smaller businesses. Many smaller operators already take social responsibility seriously: for many, the idea of signing up to some 'concordat' might appear ridiculous, as they already have strong links with their local authority through the chamber of commerce and with their local customer base. But the same may not be true for businesses located in the fringe, more disconnected from local communities, and forgotten by local authorities. These may need to be engaged. That said, in the 'global fringe', many large companies have a presence. Across the UK, High Street retailers are often found in the fringe: all the major supermarkets are represented. And then there are the shopping and leisure centres, often owned by international investment companies. If these larger companies choose to take their CSR 'obligations' more seriously in the future, then it is at least possible that more can be achieved in the fringe through public–private cooperation.

Critical choices for the rural-urban fringe

The generality of this concluding discussion is founded on a number of beliefs: first, excessive prescription is not desirable, as all fringes are different; second, there is much in England's rural-urban fringe, in terms of positive qualities, that might be emphasized, but does not need to be changed; third, planning is only one tool among many for managing the functions of the fringe; and fourth, general guidance – in the form of simple principles – is preferable to a 'blueprint' for change likely to obliterate those characteristics of the fringe – economic, social, political and environmental – that make it 'multifunctional'. Indeed, 'multifunctionality' is not something to be created in the landscape, but something to be enhanced, where necessary. We have already seen that plurality in land use and function has evolved without policy intervention. The role of planning is to learn the lessons, understand the fringe, and – where appropriate – facilitate improvements in how the landscape works. There are, of course, other reasons why we have steered clear of policy or design prescription. First, not enough is known about the nature of all land uses and functions in the fringe, their operation and their requirements. It is impossible, therefore, to prescribe solutions or to say, with any confidence, what configuration of land uses and functions might work best in the fringe. Second, the fringe – in various guises – is a global phenomenon and it is useful, therefore, to reflect on more generic responses to the needs of dynamic and diverse landscapes, rather than attempting to provide a manual for planning at the edge. Our guess is that it is only a matter of time before this appears.

We began this book with the suggestion that 'multifunctionality' is a 'conscious guide' for action, providing a framework in which to deliver sustainable development. Reflectivity is perhaps the key to delivering the right outcome: to understand context, why things are as they are, and to work with the grain rather than against it. Multifunctionality is a concept rooted in explanation, and in trying to understand interaction across functions. Sustainability, in its simplest sense, is about balanced functionality, achieving equilibrium, and allowing places to perform on several fronts simultaneously. The UK government has stated its belief that policy interventions should pursue the aims of sustainable development in 'an integrated way through a sustainable, innovative and productive economy that delivers high levels of employment, and a just society that promotes social inclusion, sustainable communities and personal well being, in ways that protect and enhance the physical environment and optimise resource and energy use' (ODPM, 2005: Para 4). In the fringe, as in any landscape, some functions work well and others less well. This is a dynamic, ever-changing, landscape: the product of thousands of years of settlement history. But very suddenly – almost overnight – it has been decided that the fringe must change: it must be planned; it must start to conform to conventional landscape tastes; it must be tidied, planted with trees, made 'safe' and sanitized; essentially, it must cease to be 'fringe'. Policy debate has turned full circle: there was heightened concern for the fringe in the 1970s, and particularly for the plight of agriculture on the edges of towns and cities. The fringe is now back in vogue. Current debate on the future of England's rural-urban fringes – a debate that has its equivalents almost everywhere in the world – presents three clear choices:

- Do nothing and continue to imagine that the fringe has a limitless capacity to absorb the detritus of the modern world;
- Seek to understand the fringe, the processes that have 'made' the landscape and that continue to make it: work with these processes, promoting its better qualities and reining back those that threaten the 'sustainability' of future development; or
- Seek to change the fringe, remoulding it in a different image; reject the processes – political, social and economic – that have made the landscape; seek to replace these with more malleable processes, with conventional planning, and ultimately create a ring of parkland – a gateway to the city within, and a bridge to the country beyond.

The first choice presents great risk, not least from urban growth and consequent sprawl. In the UK, the future of the green belt is under the spotlight. It is perhaps wise to make preparations for its replacement. Statutory containment has allowed society to forget about the fringe for the last half century, but if this containment is weakened, then some landowners and investors – those gambling on the future of the green belt – will try to cash in. By understanding the fringe – the second option – there is a chance that a framework can be put in place that will build on the best qualities of statutory containment, while also favouring the right kinds of development, with a view to improving permeability, creating economic stability, and broadening opportunities for the fringe to play a more inclusive sociocultural role. The third choice also presents great risk: by removing the 'melting pot' that the fringe represents within the wider urban system, new limits are placed on many of the functions that find a natural home at the urban edge. In a conventional landscape – a parkland of good taste – where will children build their dens or business start-ups find scruffy but cheap buildings to lease? Where will the car-breakers' yard go, the sewage be treated, or rows upon row of pre-showroom cars be stored? The fringe is an inevitable part of the city. If parts of it are sanitized and forced to comply with the latest fads in urban design, the real fringe will simply move on, re-establishing itself elsewhere. Planning faces a choice at the edge: to recognize or deny the role of fringe landscapes; to seek success by understanding and working with the processes that make the fringe, or to ignore and work against these processes. This is the context for planning on the edge.

References

Abercrombie P. (1945) *The Greater London Plan*, London, HMSO.
Abram S., Murdoch J. and Marsden T. (1997) 'The social construction of Middle England', *Journal of Rural Studies*, 12, 4, 353–364.
Adorno T. W. (1984) *Aesthetic Theory*, London, Routledge and Kegan Paul.
Advisory Council for Agriculture and Horticulture in England and Wales (1978) *Agriculture and the Countryside*, Chesham, MAFF.
Albrechts L. (2004) 'Strategic (spatial) planning re-examined', *Environment and Planning B: Planning and Design*, 31, 743–758.
Ambrose P. (1986) *Whatever Happened to Planning?*, London, Methuen (See Chapter 5: 'The war for Jenkin's ear: development in the urban fringe').
Antrop M. (1998) 'Landscape change: plan or chaos', *Landscape and Urban Planning*, 41, 155–161.
Appleton J. (1996) *The Experience of Landscape*, Chichester, Wiley.
Aragon, L. (1971) *Paris Peasant* (translated and with an introduction by Simon Watson Taylor), London, Jonathan Cape.
Archer R. W. (1973) 'Land speculation and scattered development; failures in the urban-fringe land market', *Urban Studies*, 10, 367–372.
Audirac I. (1999) 'Unsettled views about the fringe: rural-urban or urban-rural frontiers', in Furuseth O. J. and Lapping M. B. (eds) *Contested Countryside: The Rural Urban Fringe in North America*, Aldershot, Brookfield USA, Singapore, Sydney, Ashgate.
Augé M. (1995) *Non-Places: Introduction to an Anthropology of Super-Modernity*, London and New York, Verso.
Bagwell P. S. (1974) *The Transport Revolution from 1770*, London, Batsford.
Baker B. (2000) 'Full steam ahead', *Surveyor*, 17 February, 20–21.
Ball R. and Brown I. (2000) 'The industrial heritage – pressures, policies, and economic uses', *Town and Country Planning*, 69, 10, 302–303.
Barker, K. (2004) *Developing Stability: Securing our Future Housing Needs* (The Barker Review of Housing Supply), London, HM Treasury.
Barton M. (2000) *Lessons Learned: The Groundwork approach to delivering large-scale programme of brown land reclamation*, conference material for Beyond Brownfields: towards equitable and sustainable development, The Pratt Institute Centre for Community and Environmental Development, October 2000.
Bastian O. and Roder M. (1998) 'Assessment of landscape change by land evaluation of past and present situations', *Landscape and Urban Planning*, 41, 171–182.

Beauchesne A. and Bryant C. (1999) 'Agriculture and innovation in the urban fringe: the case of organic farming in Quebec, Canada', *Tijdschrift voor Economische en Sociale Geografie*, 90, 3, 320–328.

Becher B. and Becher H. (2002) *Industrial Landscapes*, Cambridge, Mass. and London, MIT Press.

Bell P., Gallent N. and Howe J. (2001) 'Re-use of small airfields', *Progress in Planning*, 55, 4, 196–260.

Benjamin W. (1999) *The Arcades Project* (translation by Eiland H. and McLaughlin K.), Massachusetts and London, Belknap Press.

BHF National Centre and Loughborough University (2004) *Health, Access to Green Space and Informal Outdoor Recreation within the Greenwood Community Forest and Nottingham City*, Cheltenham, CA.

Bianconi M., Gallent N. and Greatbatch I. (2006) 'The changing geography of sub-regional planning in England', *Environment and Planning C: Government and Policy* (forthcoming).

Bickmore Associates (2003) *The State and Potential of Agriculture in the Urban Fringe*, London, Catherine Bickmore Associates.

BK Consultants (2003) Overcoming Barriers to Better Planning and Management in the Urban Fringe, Taunton, BK Consultants.

Black Environment Network (2000) *How to Increase Ethnic Participation in National Parks and Develop a Model to Change Management Structures of Countryside Authorities and Agencies*, London, BEN.

Booth M. (1999) 'Campe-Toi! On the origins and definitions of camp', in Cleto F. (ed.) *Camp: Queer Aesthetics and the Performing Subject: A Reader*, Michigan, University of Michigan Press.

Bourassa S. C. (1991) *The Aesthetics of Landscape*, London, Belhaven.

Bovill P. (2002) 'Loosening the green belt', *Regeneration and Renewal*, 17 May, 12.

Bowman A. (2004) 'Recording industrial landscapes', *City*, 8, 3, 413–442.

Box J. and Shirley P. (1999) 'Biodiversity, brownfield sites and housing', *Town and Country Planning*, 68, 10, 306–309.

Brandt J. and Vejre H. (2003) [2000] 'Multifunctional landscapes – motives, concepts and perspectives', in Brandt J. and Vejre H. (eds) *Multifunctional Landscapes Volume 1: Theory, History and Values*, WIT Press (note: page numbers in text refer to earlier 2000 draft of this chapter).

Brandt J., Tress B. and Tress G. (eds) (2000) *Multifunctional Landscapes: Interdisciplinary Approaches to Landscape, Research and Management*, Conference material for the international conference on Multifunctional Landscapes, Centre for Landscape Research, University of Roskilde, Denmark, October 18–21, 2000.

Broughton F. (1996) 'Fringe issues', *Landscape Design*, September, 34–36.

Brownill S. (1990) *Developing London Docklands. Another Great Planning Disaster?*, London, Paul Chapman Publishing.

Bryman A. (2004) *The Disneyization of Society*, London, SAGE.

Bunker R. and Holloway D. (2001) 'Fringe city and congested countryside: population trends and policy developments around Sydney' (Urban Frontiers Program Issues Paper No. 6, January 2001), Sydney, University of Western Sydney.

Bunker R. and Houston P. (2003) 'Prospects for the rural-urban fringe in Australia: observations from a brief history of the landscapes around Sydney and Adelaide', *Australian Geographical Studies*, 41, 3, 303–323.

Burgess E. (1968) [1925] 'The growth of the city', in Park R., Burgess E. and McKenzie R., *The City*, Chicago, Illinois, University of Chicago Press.

Campaign to Protect Rural England (2001) *Green Belts: Still Working; Still Under Threat*, London, CPRE.

Campaign to Protect Rural England (2005) *Green Belt: 50 Years On*, London, CPRE.

Carlson A. (2000) *Aesthetics and the Environment: The Appreciation of Nature, Art and Architecture*, London and New York, Routledge.

Carmona M. and Gallent N. (2004) 'Planning and house building: a step change from the bottom-up?', *Town Planning Review*, 75, 1, 95–121.

Carmona M., Carmona S. and Gallent N. (2001) *Working Together: A Guide for Planners and Housing Providers*, London, T. Telford.

Carmona M., Carmona S. and Gallent N. (2003) *Delivering New Homes: Processes, Planning and Providers*, London, Routledge.

Carrión Flores C. and Irwin E. (2004) 'Determinants of residential land use conversion and sprawl at the rural-urban fringe', *American Journal of Agricultural Economics*, 86, 4, 889–904.

Carruthers J. (2003) 'Growth at the fringe: the influence of political fragmentation in United States metropolitan areas', *Regional Science*, 82, 4, 475–500.

Castells M. and Hall P. (1994) 'Technopoles: mines and foundries of the informational economy', in LeGates J. and Stout A. (eds) *The City Reader*, London and New York, Routledge.

Centre for Urban and Regional Ecology (CURE) (2002) *Sustainable Development in the Countryside Around Towns: Volumes 1 and 2*, Cheltenham, CA.

Champion T. (1985) 'Internal migration, counterurbanization and changing population distribution', in Hall, R. and White, P. (eds) *Europe's Population: Towards the Next Century*, London, UCL Press.

Chartered Institute of Housing and the Royal Town Planning Institute (2003) *Planning for Housing: The Potential for Sustainable Communities*, Coventry and London, CIH/RTPI.

Cherry G. E. and Rogers A. (1996) *Rural Change and Planning: England and Wales in the Twentieth Century*, London, E&FN Spon.

Clark W. A. V. (1987) 'The Roepke lecture in economic geography: urban restructuring from a demographic perspective', in *Economic Geography*, 63, 2, 103–125.

Clearhill, H. (1994) 'Environmentally led engineering: the Maidenhead, Windsor and Eton Flood Alleviation Scheme', Paper 45, *2nd International Conference on River Flood Hydraulics*, ed. W. R. White and J. Watts, New York, John Wiley and Sons.

Clout, H. (1972) *Rural Geography: An Introductory Survey*, London, Pergamon Press.

Coleman D. (ed.) (1990) *Europe's Population in the 1990s*, Oxford, Oxford University Press.

Collins M. (2003) 'Looking for a renaissance in country parks', *Town and Country Planning*, 72, 8, 251–253.

Congreve A. (2003) 'Ecological modernisation and rural housing', PhD thesis, King's College, London.

Counsell D. (1995) 'Can the planning system deliver Community Forests?', *Town and Country Planning*, 64, 7, 182–183.

Country Land and Business Association (CLA) (2002) *A Living Working Green Belt*, London, CLA.

Countryside Agency (1998) *State of the Countryside 1998*, Cheltenham, CA.

Countryside Agency (2002a) *Great North Forest Land Management Initiative*, Cheltenham, CA.

Countryside Agency (2002b) 'The state and potential of agriculture in the urban fringe', unpublished project brief, Cheltenham, CA.

Countryside Agency (2003) *Community Forest Programme: Evaluation Overview Report*, Cheltenham, CA.
Countryside Agency (2004) *Unlocking the Potential of Rural Urban Fringe*, Cheltenham, CA.
Countryside Agency and Groundwork Trust (2005) *The Countryside in and around Towns: A Vision for Connecting Town and Country in the Pursuit of Sustainable Development*, Cheltenham, CA.
Countryside Commission (1999) *Linking Town and Country*, Cheltenham, CC.
Cross D. F. W. (1990) *Counterurbanization in England and Wales*, Avebury, Aldershot.
Cullingworth J. B .(1996) 'A vision lost', *Town and Country Planning*, 65, 6, 172–174.
Cullingworth J. B. (1997) 'British land use planning: a failure to cope with change?', *Urban Studies*, 34, 5/6, 945–960.
Cullingworth J. B. (1999) *British Planning: 50 Years of Urban and Regional Policy*, London, Athlone Press.
Cullingworth J. B. and Nadin V. (1997) *Town and Country Planning in the UK* (12th ed.), London and New York, Routledge.
Cuthbertson A. (1996) 'Wasteland to wetlands', *Planning Week*, 13, 5.9.96, 12–13.
Daniels S. and Cosgrove D. (1988) 'Introduction: iconography and landscape', in Cosgrove D. and Daniels S. (eds) *The Iconography of Landscape: Essays on the Symbolic Representation, Design and Use of Past Environments*, Cambridge, Cambridge University Press.
Daniels T. (1999) *When City and Country Collide: Managing Growth in the Metropolitan Fringe*, Washington, DC and Covelo, California, Island Press.
Davis J. (2004) *At the Leading Edge of Sustainability: An Experimental Study of European Cities and their Environmentally Sustainable Urban Development*, Williamstown, MA, Williams College.
Department for the Environment, Food and Rural Affairs (2000) *Countryside Survey 2000: Accounting for Nature – Assessing Habitats in the UK Countryside*, London, DEFRA.
Department for the Environment, Food and Rural Affairs (2001) *The Countryside Stewardship Scheme: Traditional Farming in the Modern Environment*, London, DEFRA.
Department for the Environment, Food and Rural Affairs (2002) *Working With the Grain of Nature: A Biodiversity Strategy for England*, London, DEFRA.
Department of the Environment (1990) *Planning Policy Guidance Note Number 16: Archaeology and Planning*, London, DoE.
Department of the Environment, Transport and the Regions (1999) *Planning Policy Guidance Note 10: Planning and Waste Management*, London, HMSO.
Department of the Environment, Transport and the Regions (2000a) *Planning Policy Guidance Note 3: Housing*, London, HMSO.
Department of the Environment, Transport and the Regions (2000b) *Our Countryside: The Future: A Fair Deal for Rural England*, London, HMSO.
Department of the Environment, Transport and the Regions (2001a) *Planning Policy Guidance Note 13: Transport*, London, HMSO.
Department of the Environment, Transport, and the Regions (2001b) *Planning Policy Guidance Note 2: Green Belts*, London, DTLR.
Department of Transport, Local Government and the Regions (2000) *Planning Policy Guidance Note 11: Regional Planning*, London, DTLR.
Department of Transport, Local Government and the Regions (2001a) *Planning Policy Guidance Note 7: The Countryside*, London, DTLR.

Department of Transport, Local Government and the Regions (2001b) *Planning Policy Guidance Note 4: Industrial and Commercial Development and Small Firms*, London, DTLR.

Department of Transport, Local Government and the Regions (2002) *Planning Policy Guidance Note 9: Nature Conservation*, London, DTLR.

Dewar D. (2002) 'Is it time to loosen the belt?', *Planning*, 1470, 24 May, 8.

Douglas I. and Box J. (2000) 'The changing relationship between cities and biosphere reserves': a report prepared for the Urban Forum of the United Kingdom Man and Biosphere Committee (derived from a workshop held in Manchester in 1994), UNESCO.

Douglass H. (1925) *The Suburban Trend*, New York, Century Co.

Downey P. (1980) *The Impact of Brent Cross*, London, Greater London Council.

EcoRegen (2005) *Some Examples*, http://www.ecoregen.org.uk/home/introduction/example.html (accessed 20 June 2005).

Edwards B. (2000) 'Design guidelines for sustainable housing', in Edwards B. and Turrent D. (eds) *Sustainable Housing: Principles and Practice*, London, E&FN Spon.

Edwards M. (2000) 'Sacred cow or sacrificial lamb? Will London's green belt have to go?', *City*, 4, 1, 105–112.

Egan, Sir John (2004) *The Egan Review: Skills for Sustainable Communities*, London, ODPM.

Elson M. J. (1986) *Green Belts: Conflict Mediation in the Urban Fringe*, London, Heinemann.

Elson M. J. (2003) 'A "take and give" green belt?', *Town and Country Planning*, 72, 3, 104–105.

England Rural Development Programme (ERDP (2003) *Annual Report to the European Commission* (available at http://www.defra.gov.uk/erdp/docs/nwchapter/nwintroview.htm; accessed 3 February 2004).

English Heritage (2004a) *Farming the Historical Landscape: Caring for Archaeological Sites in Grassland*, London, EH.

English Heritage (2004b) *Farming the Historical Landscape: Caring for Historic Parkland*, London, EH.

Entec UK (2003a) *REACT Initiative Monitoring and Evaluation*, Cheshire, Entec UK Limited.

Entec UK (2003b) *Urban Fringe Action Plans: Their Role, Scope and Mechanisms for Implementation*, Cheshire, Entec UK Limited.

European Greenways Association (2000) *The European Greenways: Good Practice Guide: Examples of Actions undertaken in Cities and Periphery*, (A.E.V.V./E.G.W.A.), Brussels, EC Directorate-General for the Environment.

European Multi-Stakeholders Forum on CSR (2004) *Corporate Social Responsibility: Final Results and Recommendations*, Conference proceeding (available at http://forum.europa.eu.int; accessed 15 June 2005).

Evans K., Taylor I. and Fraser P. (2000) 'Shop 'til you drop', in Miles M., Hall T. and Borden I. (eds) *The City Cultures Reader*, London and New York, Routledge.

Fairbrother N. (1970) *New Lives, New Landscapes*, London, Architectural Press.

Fishman R. (1996) 'Beyond suburbia: the rise of the technoburb', in LeGates J. and Stout A. (eds) *The City Reader*, London and New York, Routledge.

Foot J. (2000) 'The urban periphery, myth and reality: Milan, 1950–1990', *City*, 4, 1, 7–26.

Freidberger M. (2000) 'The rural-urban fringe in the late twentieth century', *Agricultural History*, 74, 2, 502–514.

Fyson A. (2002) 'Green belt revisionists', *Planning*, 1469, 17 May, 10.
Gallent N. and Carmona M. (2004) 'Planning for housing: unravelling the frictions in local practice', *Planning Practice & Research*, 19, 2, 123–146.
Gallent N., Andersson J. and Bianconi M. (2004) *Vision for a Sustainable, Multi-Functional Rural-Urban Fringe*, Cheltenham, CA.
Gallent N., Shoard M., Andersson J. and Oades R. (2003a) *Urban Fringe – Policy, Regulatory and Literature Research: Final Report*, Cheltenham, CA.
Gallent N., Shoard M., Andersson J. and Oades R. (2003b) *Policy, Regulatory and Literature Research: Nature Conservation, History and Archaeology*, Cheltenham, CA.
Garreau J. (1991) *Edge City: Life on the New Frontier*, New York, Doubleday.
Gilbert O. (1989) *The Ecology of Urban Habitats*, London, Chapman and Hall.
Gilg A. G. (1985) *An Introduction to Rural Geography*, London, Edward Arnold.
Gold J. R. and Gold M. M. (1990) 'A Place of Delightful Prospects': promotional imagery and the selling of suburbia', in Zonn L. (ed.) *Place Images in Media: Portrayal, Experience and Meaning*, Savage, MD, Rowman and Littlefield.
Goode D. (1997) 'The nature of cities', *Landscape Design*, September, 14–16.
Government Office for the North West (2003) *Regional Planning Guidance 13: North West, RPG Partial Review 2003*, London, TSO.
Greater London Authority (2002) *Connecting with London's Nature: The Mayor's Biodiversity Strategy*, London, GLA.
Greed C. H. (ed.) (1999) *Social Town Planning*, London, Routledge.
Griffiths J. (1994) 'The last frontier', *Planning Week*, 2, 11, 14–15.
Gummer, J. (2002) 'Concreting over the countryside', *Planning*, 1469, 17 May, 9.
Guy C. M. (1998) '"High street" retailing in off-centre retail parks: a review of the effectiveness of land use planning policies', *Town Planning Review*, 69, 3, 291–313.
Gwilliam M., Bourne C., Swain C. and Pratt A. (1998) *Sustainable Renewal of Suburban Areas*, York, JRF.
Hall P. (1989) *London 2001*, London, Unwin Hyman.
Hall P. (1994) 'Abercrombie's Plan for London, 50 years on: A Vision for the Future', Report of the Second annual Vision for London Lecture, London, Vision for London.
Hall P. (1998) *Cities in Civilization: Culture, Innovation, and Urban Order*, London, Phoenix Giant.
Hall P., Thomas R., Gracey H. and Drewett R. (1973) *The Containment of Urban England: Urban and Metropolitan Growth Processes or Megalopolis Denied*, London, Sage Publications (Volume 1).
Hall R. (1985) 'Households, families and fertility', in Hall R. and White P. (eds) *Europe's Population: Towards the Next Century*, London, UCL Press.
Harris R. and Lewis R. (1998) 'Constructing a fault(y) zone: misrepresentations of American cities and suburbs, 1900–1950', *Annals of the Association of American Geographers*, 88, 4, 622–639.
Harrison C., Burgess J., Millwood A. and Dawe G. (1996) *Accessible Natural Greenspace in Towns and Cities: A Review of Appropriate Size and Distance Criteria*, Peterborough, English Nature.
Hasler B. and Kjellerup U. (2000) Second Draft of 'Recommendations on interdisciplinary landscape research – Workshop No. 4: Complexity of landscape management', in Brandt J., Tress B. and Tress G. (eds) *Multifunctional Landscapes: Interdisciplinary Approaches to Landscape, Research and Management*, Conference material for the international conference on Multifunctional Landscapes, Centre for Landscape Research, University of Roskilde, Denmark, 18–21 October 2000.

Healey P. (1997) 'An institutionalist approach to spatial planning', in Healey P., Khakee A., Motte A. and Needham B. (eds) *Making Strategic Spatial Plans: Innovation in Europe*, London, UCL Press.

Heimlich R. and Anderson D. (2001) 'Development at the urban fringe and beyond: impacts on agriculture and rural land', *Agricultural Report*, AER 803, 88.

Highways Agency (2003) *Biodiversity Action Plan*, London, Highways Agency.

Hite, J. (1998) *Land Use Conflicts on the Urban Fringe: Causes and Potential Resolution*, Strom Thurmond Institute, Clemson University, South Carolina.

HM Government (2000) *Our Countryside – The Future: A Fair Deal for Rural England*, London, TSO.

HM Government (2002) *The Landfill (England and Wales) Regulations 2002*, London, HMSO.

Hoggart K. (2003) 'England', in Gallent N., Shucksmith M. and Tewdwr-Jones M. (eds) *Housing in the European Countryside: Rural Pressure and Policy in Western Europe*, London, Routledge.

Hoskins W. G. (1955) *The Making of the English Landscape*, London, Hodder and Stoughton.

House of Commons Environment, Transport and Regional Affairs Select Committee (1999) *Town and Country Parks*, London, TSO.

House of Commons Environment, Transport and Regional Affairs Committee (2000) *Twentieth Report: UK Biodiversity – Interim Report*, London, TSO.

Howard E. (2003) [1898] *To-morrow: A Peaceful Path to Real Reform* (Original Edition with New Commentary by Peter Hall, Dennis Hardy and Colin Ward), London, Routledge.

Howard E. B. and Davies R. L. (1993) 'The impact of regional out of town retail centres: the case of the Metro Centre', *Progress in Planning*, 40, 2, 89–165.

Ilbery B. W. (1991) 'Farm diversification as an adjustment strategy on the urban fringe of the West Midlands', *Journal of Rural Studies*, 7, 3, 207–218.

Jackson P. (1999) 'Consumption and identity: the cultural politics of shopping', *European Planning Studies*, 7, 1, 25–39.

Jenks, M. and Burgess, R. (eds) (2000) *Compact Cities: Sustainable Urban Forms for Developing Countries*, London: Spon Press.

Johnston B. (2000) 'Industry still prefers living on the edge', *Planning*, 25 February, 14–15.

Jones P. and Vignali C. (1994) 'Factory outlet shopping centres – developments and planning issues', *Housing and Planning Review*, February–March, 9–11.

Kaika M. and Swyngedouw E. (2000) 'Fetishizing the modern city: the Phantasmagoria of urban technological networks', *International Journal of Urban and Regional Research*, 24, 1, 120–138.

Kampfner J. (2003) 'Let's revive the regions instead of destroying the green belt: save this country from urban sprawl', *The Express*, 30 April, 12.

Kamvasinou K. (2003) 'The poetics of the ordinary: ambiance in the moving transitional landscape', in Dorrian M. and Rose G. (eds) *Deterritorialisations. . . . Revisioning Landscapes and Politics*, London and New York, Black Dog Publishing Limited.

Kendle T. and Forbes S. (1997) *Urban Nature Conservation: Landscape Management in the Urban Countryside*, London, Taylor and Francis.

Kunzmann K. (2000) 'Strategic spatial development through information and communication', in Salet W. and Faludi A. (eds) *The Revival of Strategic Spatial Planning*, Amsterdam, Royal Netherlands Academy of Arts and Sciences.

Land Use Consultants (2001) 'The Green Arc Partnership: a prospectus', unpublished, London, Land Use Consultants.

Land Use Consultants (2002) *The Landscape Policies and Practices of Major National Institutions, Volumes One and Two*, Cheltenham, CA.
Land Use Consultants (2003) *Planning for Sustainable Rural Economic Development, Part A: Position Statement and Supporting Report*, Bristol, LUC.
Land Use Consultants (2004) *Bringing the Big Outdoors Closer to People: Improving the Countryside around London – The Green Arc Approach*, London, LUC.
Landscape Design Associates (2004a) *New Trees, New Landscapes: Recommendations for Future Tree Planting Strategies for the Watling Chase Community Forest*, Herefordshire County Council.
Landscape Design Associates (2004b) *New Trees, New Landscapes: A Review of Tree Planting Strategies for the Watling Chase Community Forest*, Herefordshire County Council.
Leach R. and Percy-Smith J. (2001) *Local Governance in Britain*, Basingstoke, Palgrave.
Leedale M. (2001) 'A very sticky issue of green gap constraint', *Planning*, 1413, 6 April, 8.
Ling C. (2001) *Rebuilding the Post-industrial Landscape: Interaction between Landscape and Biodiversity on Derelict Land*, on-line resource 'Toolkits for Community Led Regeneration of Derelict Land', Manchester, Centre for Urban and Regional Ecology.
Ling C., Handley J. and Rodwell J. (2000) 'Multifunctionality and scale in post-industrial land regeneration', paper presented to the International Conference on Multi-functional Landscapes, Centre for Landscape Research, University of Roskilde, Denmark, 18–21 October 2000.
Lober T. and Gallent N. (2005) 'Planning's failures from a general aviation perspective', paper presented at the Planning Research 2005 Conference, University of Manchester, March 2005.
Lobley M., Errington A. and McGeorge A. et al. (2002) *Implications of Changes in the Structure of Agricultural Businesses*, London, DEFRA.
Lockwood J. (2001) 'Can we curb the "carcentrics" culture?', *Planning*, 30 March, 12–13.
Lowe M. S. (2000) 'Britain's regional shopping centres: new urban forms?', *Urban Studies*, 37, 2, 261–274.
Lowenthal D. and Prince H. (1965) 'English landscape tastes', *Geographical Review*, 55, 2, 185–222.
Lynch, K. (1960) *The Image of the City*, Cambridge, Mass., London, MIT Press.
Matless D. (1998) *Landscape and Englishness*, London, Reaktion Books.
Ministry of Agriculture, Fisheries and Food (2000) *England Rural Development Programme 2000–2006, North West Region, Appendix A2*, http://www.defra.gov.uk/docs/ nwchapter/nwintroview.htm; accessed 3 February 2004.
Morgan P. (2002) 'Changing places – going soft on derelict land', *ECOS*, 22, 3/4, 4, Northampton, BANC.
Mori H. (1998) 'Land conversion at the urban fringe: a comparative study of Japan, Britain and the Netherlands', *Urban Studies*, 35, 9, 541–558.
Morris H. (2001) 'Green belt mind set needs an overhaul', *Planning*, 1404, 2 February, 3.
Muir R. (1999) *Approaches to Landscape*, London, Macmillan Press.
Muir R. (2003) 'On change in the landscape', *Landscape Research*, 28, 4, 383–403.
Murdoch J. and Abram S. (2002) *Rationalities of Planning: Development Versus Environment in Planning for Housing*, Basingstoke, Ashgate.
Nairn, I. (1955) *Outrage*, London, Architectural Press.

Nairn, I. (1957) *Counter-Attack against Subtopia*, London, Architectural Press.
Newchurch and Co. (1999) *A Working Definition of Local Authority Partnerships*, London, DETR.
Nkambwe M. (2003) 'Contrasting land tenures: subsistence agriculture versus urban expansion on the rural-urban fringe of Gaborone, Botswana', *International Development Planning Review*, 25, 4, 391–405.
Office of the Deputy Prime Minister (2001) *Planning Policy Statement 7: The Countryside – Environmental Quality and Social Development*, London, ODPM.
Office of the Deputy Prime Minister (2002a) *Valuing the external benefits of undeveloped land: a review of the economic literature*, London, ODPM.
Office of the Deputy Prime Minister (2002b) *Planning Policy Guidance Note 17: Planning for Open Space, Sport and Recreation*, London, ODPM.
Office of the Deputy Prime Minister (2002c) *Strategic Gap and Green Wedge Policies in Structure Plans*, London, ODPM.
Office of the Deputy Prime Minister (2003a) *Sustainable Communities: Building for the Future*, London, ODPM.
Office of the Deputy Prime Minister (2003b) *Living Places: Cleaner, Safer, Greener*, London, ODPM.
Office of the Deputy Prime Minister (2003c) *Contributing to Sustainable Communities – A New Approach to Planning Obligations*, London, ODPM.
Office of the Deputy Prime Minister (2003d) *Survey of Urban Design Skills in Local Government*, London, ODPM.
Office of the Deputy Prime Minister (2004a) *Planning Policy Statement 6: Planning for Town Centres*, London, ODPM.
Office of the Deputy Prime Minister (2004b) *Creating Sustainable Communities: Greening the Gateway*, London, ODPM.
Office of the Deputy Prime Minister (2004c) *Planning Policy Statement 7: Sustainable Development in Rural Areas*, London, ODPM.
Office of the Deputy Prime Minister (2005) *Planning Policy Statement 1: Delivering Sustainable Development*, London, ODPM.
O'Neill J. (2003) 'The future of the green belt: what are the options?', *Planning*, 1509, 7 May.
Orwell G. (1939) *Coming Up for Air*, London, Penguin Books.
Osment F. (1998) 'The development–countryside conflict', *Landscape Design*, 274, October.
Oxford Brookes University (2002) *Planning for Delivery of Quality of Life Benefits*, Oxford Brookes University: Oxford.
Pahl R. (1965) *Urbs in Rure: The Metropolitan Fringe in Hertfordshire*, London, London School of Economics.
Palen J. (1995) *The Suburbs*, New York, McGraw Hill.
Parkhurst G. (2000) 'Influence of bus-based park and ride facilities on users' car traffic', *Transport Policy*, 7, 2, 159–172.
Pearman H. (1991) 'Business parks', *Architectural Review*, 189, 1129, 61.
Penn Associates (2002) 'The National Community Forest Partnership: examples of good practice', unpublished, Hexham, Penn Associates.
Penn Associates (2003) 'The economic impact of the Community Forest Programme', unpublished, Hexham, Penn Associates.
Picon A. (2000) 'Anxious landscapes: from the ruin to rust', *Grey Room*, 1, 1, 64–83, Cambridge Mass., MIT Press.
Planning (2005) 'Poll shows location as a key home factor', *Planning Magazine*, 10 June, 4.

Preston C. D., Telfer M. G., Arnold H. R., Carey P. D., Cooper J. M., Dines T. D., Hill M. O., Pearman D. A., Roy D. B. and Smart S. M. (2002) *The Changing Flora of the UK*, London, DEFRA.
Priemus H., Rodenburg C. H. and Nijkamp P. (2004) 'Multifunctional urban land use: a new phenomenon? A new planning challenge?', *Built Environment*, 30, 4, 269–273.
Pryor R. (1968) 'Defining the rural-urban fringe', *Social Forces*, 47, 202–215.
Punter J. V. (1985) *A History of Aesthetic Control 1947–1985, Part Two*, Working Papers in Land Management and Development, Reading, University of Reading.
ReUrbA (2001) 'Urban fringe – an introduction', paper available for download at www.reurba.org
Roberts B. K. (1987) *The Making of the English Village: A Study in Historical Geography*, Harlow, Longman Scientific & Technical.
Rodenburg C. H. and Nijkamp P. (2004) 'Multifunctional land use in the city: a typological overview', *Built Environment*, 30, 4, 274–288.
Rowe P. G. (1991) *Making a Middle Landscape*, Cambridge, Mass., MIT Press.
RICS (Royal Institute of Chartered Surveyors) (2000) *Transport Development Areas*, London, RICS.
Royal Town Planning Institute (2000) *Green Belt Policy: A Discussion Paper*, September, London, RTPI.
Royal Town Planning Institute (2002) *Modernising Green Belts: Discussion Paper*, May, London, RTPI.
Royal Town Planning Institute (2003) *A Manifesto for Planning*, London, RTPI.
Royal Town Planning Institute (2004) *Response to 'Unlocking the Potential' Consultation*, London, RTPI.
RSKENSR (2003) *Business Case for the Environment Research Study: Summary of Principal Findings*, Helsby, Greening for Growth.
RSKENSR (2004) *Greening for Growth: Delivering a Qualitiy Environment for People and Business* (Business Case for the Environment Research Study: Report 2: Strategy to Attract Investors), Helsby, Greening for Growth.
Rydin Y. and Myerson G. (1989) 'Explaining and interpreting ideological effects: a rhetorical approach to green belts', *Environment and Planning D: Society and Space*, 7, 463–479.
Samuel R. (2000) 'Theme parks – why not?', in Miles M., Hall T. and Borden I. (eds) *The City Cultures Reader*, London and New York, Routledge.
Shoard M. (1979) 'Metropolitan escape routes', *London Journal*, 5, 1, 87–112.
Shoard M. (1980) *The Theft of the Countryside*, London, Temple Smith.
Shoard M. (1997) *This Land is Our Land: The Struggle for Britain's Countryside*, London, Gaia Books.
Shoard M. (1999) *A Right to Roam: Should We Open up Britain's Countryside?*, London, Oxford University Press.
Shoard M. (2000) 'Edgelands of promise', *Landscapes*, 1, 2, 74–93.
Shoard M. (2002) 'Edgelands', in Jenkins J. (ed.) *Remaking the Landscape: The Changing Face of Britain*, London, Profile Books.
Sieverts T. (2003) *Cities without Cities: An Interpretation of the Zwischenstadt*, London and New York, Spon Press.
Sinclair I. (2002) *London Orbital: A Walk Around the M25*, London, Granta Books.
Skellern, J. (1993) 'Countryside recreation in Berkshire: a study asking whether current trends in countryside recreation in Berkshire are towards the reinforcement of an exclusive countryside', unpublished dissertation, MSc Town and Country Planning, University of Reading.

Smith N. (2001) 'Green belt policy in need of update for public spaces', *Planning*, 1419, 18 May, 7.

South East England Regional Assembly (SEERA) (2005) *A Clear Vision for the South East: South East Plan Consultation Draft*, Guildford, SEERA.

Sport England with the Countryside Agency and English Heritage (2003) *The Use of Public Parks in England*, London, Sport England.

Stathis D. (1986) 'The historic evolution of the motorway programme in Britain and its socio-economic effects', unpublished MSc Thesis, UCL.

Statistics Canada (2001) *2001 Census Dictionary*, available for download at http://www.statcan.ca/english/census2001/dict/geo050.htm.

Sullivan W. C. (1996) 'Cluster housing at the rural-urban fringe: the search for adequate and satisfying places to live', *Journal of Architectural and Planning Research*, 13, 4, 291–309.

Tandy C. (1975) *Landscape of Industry*, London, Leonard Hill Books.

Taylor A. F., Kuo F. E. and Sullivan W. C. (2001) 'Coping with ADD: the surprising connection to green play settings', *Environment and Behaviour*, 33, 1, 54–77.

TEP and Vision 21 (2004) *Newlands Community Involvement Research Volumes 1 and 2*, TEP and Vision 21.

Tewdwr-Jones M. (1997) 'Green belts or green wedges for Wales: a flexible approach to planning in the urban periphery', *Regional Studies*, 31, 1, 73–77.

Tewdwr-Jones M. (2004) 'Spatial planning: principles, practices and cultures', *Journal of Planning & Environmental Law*, May (B19–B20), 560–569.

Thayer R. (2002) 'Three dimensions of meaning', in Swaffield S. (ed.) *Theory in Landscape Architecture: A Reader*, Philadelphia, University of Pennsylvania Press.

Thomas D. (1970) *London's Green Belt*, London, Faber.

Thomas D. (1990) 'The edge of the city', *Transactions of the Institute of British Geographers*, New Series, 17, 2, 131–138.

Town and Country Planning Association (2002) *'Green Belts' Policy Statement*, London, Town and Country Planning Association.

Trainer T. (1998) 'Towards a checklist for eco-village development', *Local Environment*, 3, 1, 79–83.

Travers T. (2002) 'Is it time to think the unthinkable on the green belt?', *Evening Standard*, 9 July, 11.

Trinder B. (1997) *The Making of the Industrial Landscape*, London, Phoenix Giant.

Urban Task Force (1999) *Towards an Urban Renaissance*, London, Spon Press.

Vincent and Gorbing Ltd (2004) *Community Green Space and New Development: Creation through the Planning System and Lessons for the Future*, Cheltenham, Hertfordshire Countryside Management Service, Watling Chase Community Forest and the Countryside Agency.

Voigt W. (1996) 'From the hippodrome to the aerodrome, from the air station to the terminal: European airports, 1909–1945', in Zukowsky J. (ed.) *Building for Air Travel: Architecture and Design for Commercial Aviation*, Munich and New York, Prestel.

Ward S. (2004) *Planning and Urban Change*, second edition, London, SAGE.

Wehrwein G. S. (1942) 'The rural urban fringe', *Economic Geography*, 18, 3, 217–228.

Westhead, P. and Batstone, S. (1998) 'Independent technology-based firms: the perceived benefits of a science park location', *Urban Studies*, 35, 12, 2197–2219.

Whitehand J. W. R. (1967) 'Fringe belts: a neglected aspect of urban geography', *Transactions of the Institute of British Geographers*, 41, June, 223–233.

Whitehand J. W. R. (1988) 'Urban fringe belts: development of an idea', *Planning Perspectives*, 3, 1, 47–58.

Whitehand J. W. R. (1992) *The Making of the Urban Landscape*, Institute of British Geographers Special Publication 26, Oxford, Basil Blackwell.

Whitehand J. W. R. and Morton N. (2003) 'Fringe belts and the recycling of urban land: an academic concept and planning practice', *Environment and Planning B: Planning and Design*, 30, 6, 819–839.

Whitehead, C. (2003) 'Interview material', in *Urban Regeneration: The New Agenda for British Housing, Creating new Communities*, London, Building for Life and English Partnerships.

Wibberley G. (1959) *Agriculture and Urban Growth: a study of the competition for rural land*, London, Joseph.

Wilce H. (2004) 'Lesson's in nature's classroom', *Independent Education Supplement*, 15 July.

Williams J. and Brown N. (eds) (1999) *An Archaeological Research Framework for the Greater Thames Estuary*, Essex County Council, Kent County Council, English Heritage and the Thames Estuary Partnership.

Williams K., Burton E. and Jenks M. (eds) (2000) *Achieving Sustainable Urban Form*, London, E&FN Spon.

Williams-Ellis C. (1975) [1928] *England and the Octopus*, Glasgow, Blackie.

Willis S. and Holliss B. R. (1987) *Military Airfields of the British Isles*, Playa del Rey, California, Enthusiasts Publications.

Winter J. (2005) 'All the world's car park', *Guardian*, 25 January.

Wood R. and Ravetz J. (2000) 'Recasting the urban fringe', *Landscape Design*, 294, 10, 13–17.

Index

Illustrations are indicated by a page number in **bold**

'A5 Corridor Initiative' 82
Abercrombie, Patrick 34, 36, 164
Aberystwyth 52
Abram, S. 140
acceptability, political 38, 39
access to the fringe 93, 116, 131–2, 137, 139, 206–8; to common land/woodland 127; economic and social barriers 187, 207–8; illegal 133; importance of, in new economy 93; physical 133–5; for recreation 67, 68, 131–6; to reservoirs 127–8; to riverbanks 127–8
access to road networks 5, 6, 93
Action Plans 196, 197
Adorno, T. 77, 81, 84
Advisory Council for Agriculture and Horticulture 105–6
aesthetic fringe 77, 117, 123, 178
aesthetic functionality 21
aesthetics 38, 143; of urban dowry 85–6
agriculture *see* farming
airfields 59, **61,** 120, 125; re-use of 24, 49, 50, 59–61, 93
airports 43, 125
Albrechts, L. 166, 180, 182
allotments 73, 116
Ambrose, P. 168
Antrop, M. 204
Appleton, J. 81–2
Aragon, Louis 85–6
archaeology: planning and 53–4; pre-modern 52–5
Archer, R.W. 35

architecture of fringe: regional 81–2; standardization of 81, 96
army barracks 61
asylum reception centres 6, 38, 75
Audirac, I. 7
Augé, M. 18, 44, 78
Australia 10, 12

Barker Review 141
Barn Elms 153–5, 205
Bath 82, 107
beauty 81–2; changing notions of 85–6
Becher, B. and H. 80, 85
BedZED development 121–2, **122**
Belfast 145
Bell, P. 24
Benjamin, W. 79
Berkshire, recreation provision 68
Big Pit 50
biodiversity 142, 143, 145, 146
Biodiversity Action Plans 146, 147
Birmingham, canals in 50, 68
Bluewater shopping centre 79, 102, 124
Bold Moss, St Helens 22
Booth, M. 89
boundaries, between urban and rural areas 34, 92, 164, 175
Bovill, P. 165
Bowman, A. 85
Box, J. 141, 143
Brandt, J. 20
Brent Cross shopping centre 40, 100
Bristol Parkway **98**, 98, 99

brown-field sites 59, 119–20, 142, 143, 168
Bunker, R. 10, 12
business parks 4, 15, 43, 93, 102, 103, 133

Campaign to Protect Rural England (CPRE) 34, 167, 168, 172–3, 174
canals 49–50, **126,** 127–8; walking along 67
car: dependency on 39, 40, 99, 100, 101, 135–6; and making of the fringe 31, 43, 45
car-breaking 4, 73, 75
car parks **97,** 98–9
caravan sites 6, 73
Carlson, A. 88
Carmona, M. 187
Carruthers, J. 12, 21–2, 33
Castells, M. 43–4
catteries/kennels 6, 73
Centre for Urban and Regional Ecology 10, 188, 190, 202
Chartered Institute of Housing 124
Chelmsford, archaeology of the fringe 52
children: opportunities for education in the fringe 129, 130; play in the fringe 144
cluster housing 120, 121, 122, 123, 125, 137, 210
Cohen, Jem, *Chain* 80
Colchester, archaeology in the fringe 53
Colliers Moss Common 22–3, **23,** 24, 50, 203
commercial leisure facilities 6, 14, 73, 75, 128; and informal recreation 136, 137
Common Agricultural Policy 143
common land 127, 128
'Communities Plan' 119
community 36–8, 121, 124, 137
Community Forests 16, 185, 193–4, 195; environmental regeneration 130, 147, 191; health benefits of 130–1; improving the landscape 83–4; partnership working and 146, 147, 183–4, 204–5
community of users 115, 125, 131, 137; participation of 188, 189, 192–3
commuter estates 118, 120–1, 124–5, 137, 198
commuters 30, 98, 125
compulsory purchase orders 194
Congreve, A. 121
conservation 15, 50, 55, 66; of historic buildings 50; tension with development 12, 35–6, 59, 109, 139–40, 154–5; through re-use 55, 57–62, 63, 66
consumption of countryside 42, 43, 73
containment (*see also* green belt) 35, 170, 172, 178, 210, 213; objectives of 161
corporate social responsibility 151, 152, 194, 205, 211
Cosgrove, D. 71–2
cotton mills 49, 55
Country Land and Business Association, and green belt reform 17
Country Landowners Association 178
country parkland 8, 14, 66–7, 74, 126, 127, 132
Countryside Agency 126, 136, 190, 208; and Community Forests 83, 193–4; definition of fringe 5; REACT programme 191; vision for the future 68–9, 93, 195, 198
Countryside and Rights of Way Act 2000 132
Countryside Commission 126, 156
Countryside Stewardship 146, 148
Cross, D. 31
CPRE *see* Campaign to Protect Rural England
Cullingworth, J.B. 162–3
cycling/cycle-ways 134, **134,** 207

Daniels, S. 71–2
Davis, J. 122
decentralization 31; of industry 29; of population 30, 73, 92; of retail 93, 99–100
definition of fringe 4, 5
DEFRA (Department for the Environment, Food and Rural Affairs) 147, 148
derelict land 42, 57, 69, 83, 128; regeneration of 21–2
design 38, 40, 43,180; of housing 120–1, 123–4
development (*see also* housing) 4, 12, 14–15, 36, 96–7; aesthetic quality of 38; balance with ecology 140–4, 149, 151–2, 154–5; control of 162–3 (in green belts 174, 176, 177); local authorities and 172; need for 177; opposition to 37–8; pressure on industrial heritage 58, 59; tension with conservation/ protection 12, 35–6, 59, 109, 139–40, 154–5

228 *Index*

development fingers 74, 172
Dewar, D. 167
distribution centres 4, 5, 15, 104, 149
districts 206
Dorset 145
Dunbar 145

eco-parks 176, 178
'eco-villages' 120, 121
ecological fringe 138–40, 145–57, 179; balance with development 140–4, 149, 151–2, 154–5; and recreation 207
ecological functionality 21, 22, 192
ecological modernization 120, 121–2
economic fringe 91–4, 178
economic functionality 21, 203
economic restructuring 38, 40–1, 45, 91–4
Ecup 67, 127
edge cities 102–5, 113, 118, 137
edges 206
educational benefits 129–30,
educational institutions, in the fringe 6, 73
Edwardian fringe belts 12
Edwards, B. 124
Edwards, M. 167, 168
Elson, M.J. 4, 167
energy production and distribution 14, **89**
English Nature 145
environment, and housing development 120–2, 123–4
Environment Agency 148–9
environmental enhancement 174–5
environmental regeneration 130, 147, 191, 204
Epping Forest 8, 127, 135
equestrian centres 8, 74
ethnic minority groups, access to fringe 208
'eyesore argument' 88

Fairbrother, N. 86
farm buildings, conversion to business use 33
farm shops 8, 107
farming 62, 107–9, **108**, 176; changes in 41–2, 105; conflict with other land uses 7–8, 12, 105–6, 108, 114, 172, 197; diversification of 16, 33, 42, 75, 107–9, 170, 174, 177; environmental damage of 142–3; fragmentation of farms 1, 106; multifunctionality in 19–20; planning

and 178, 196–7; proximity to urban markets 92, 105, 169
farmland 3, 7–8, 16, 73, 75; urban encroachment on 7, 12
films, representation of fringe in 18, 80
fishing lakes 6
Fishman, R. 102
flood alleviation schemes 149
fly-tipping 7, 74, 106, **106**, 114, 16
food miles 105, 107, 114
footpaths 132, 133, 149, 207
Forbes, S. 146
forestry (*see also* Community Forests) 16, 176
Freidberger, M. 33
fringe *see* rural-urban fringe
fringe belt 31
functionality (*see also* multifunctionality) 17, 19, 69, 76–7, 88
Fyson, A. 167

Gallent, N. 187
garden centres 6, 14, 15, 73, 75
Garreau, J. 102–3
Germany, Duisberg-Nord 57–8
golf courses 6, 14, 73, **117**
Goode, D. 156
Grand Union Canal 49–50
grasslands, unimproved 115, 125, 138
gravel pits 67–8, 128
green activities 125, 178, 182
green belts 3, 27, 34, 142, 164–5; alternatives 169–73, 210; calls for reform 167–8, 169, 199; drawbacks of 178–9; effectiveness of 165; housing on 15, 34, 167; and land-use planning 166; London's 126, 135, 165, 167; modernization 173–9; pressure on 27, 35–6; sale of land 167, 210; support for 168–9
green wedges 74, 165, 170, 171, 172, 177, 206
Greening for Growth 192
'Greening the Gateway' 129–30, 199
Greenways 68, 133–4, 136, 207
Griffiths, J. 4, 179
Groundwork Trust 57, 65, 83, 136
Gummer, John 167
Guy, C.M. 39

Hall, P. 30, 34, 43–4, 80, 92
Hanningfield Reservoir 67, 127–8
Hasler. B. 204
health benefits of the fringe 130–1, 192
Heathrow Airport 125
hedgerows, protection of 63, 64

Hertsmere Borough, Community Forest partnership 190
Highways Agency, Environmental Strategic Plan 151–2
historic environment 15, 38, 49, 178; of the fringe 50, 52–3, 64, 66–9
historical functionality 21
Hockerton Energy Project 123–4
Hoggart, K. 164
Holloway, D. 10, 12
'horsiculture' 74, 108
horticulture 8, 11
Hoskins, W.G., *The Making of the English Landscape* 28–9, 77–8
housing 15, 34, 102, 115, 118–20, **119,** 141; alongside retail centres 115, 118, commuter 118, 120–1, 124–5, 137, 198; design issues 120–1, 123–4; environmentally friendly 121–4; on green belt land 15, 34, 167–8; in South East England 59, 141, 164
Howard, Ebenezer 34, 137, 164
Howdon Wetlands 153, 155, 156, 203

illegal activities 3, 7
inclusion 181, 183, 202; and access to fringe 187
industrial heritage 50, **51,** 55, **56,** 59; recreational potential 67; and re-use 55, 57–82; theme parks 58, 69
industrial landscape 57; aesthetics of 85–6
Industrial Revolution 31, 62, 91
industry *see* light industry; manufacturing industry
infill development 141
informal recreation 14, 125–6, 128–9, 144; access to fringe 67, 68, 127–8, 131–6; and commercial leisure 136, 137; on historic fringe 66–9; on landfill sites 65–6, 111
inner cities, need to revitalize 119
integrated management 181, 182, 186–7, 202; partnership working and 188, 190–2
integration of functions 11, 20–1, 111–12, 117, 136, 153, 154, 188, 189, 203–4
interaction between functions 181, 183, 204–5
Ipswich Airfield 60

Johnston, B. 102
Jubilee River 13, 111, 149

Kaika, M. 85–6
Kampfner, J. 167–8
Kamvasinou, K. 30
Kendle, T. 146
Kent Thames Gateway 54, 111, 199
Kjellerup, U. 204

lakes 73, 127
land speculation 35, 73–4, 210
land-use categories 10–11, 13–16, 73–4, 75
land-use planning 39, 142, 162–4, 180, 182, 196; environmental planning 155; and green belt policy 164–9; and multifunctionality 20–2, 37, 183; neglect of fringe 183, 185; and out-of-town shopping 39–40
landfill sites 64, 65–6, 112
landmarks 206
landscape 28, 69, 71, 143; changing perceptions of 85–90; community and 36–8; definition of 71–2; improving 81–4; making of 28, 29, 44–5; physical 72–6; politics and 38–40, 45; remaking of 63, 69; unimproved 128–9
landscape improvement schemes 82–3, 90
'lawlessness' 18, 25–6, 80, 81
leadership 181, 186, 188, 193–6, 202
Lee Valley Regional Park 68, 133
Leedale, M. 168
leisure centres 128
light industry 92, 93, 97; relocation of 91
Ling, C. 21
liveability agenda 129, 132, 199, 208
Liverpool 96
local authorities 33–4, 143, 147, 148, 162, 179; economic development in the fringe 172; and local partnerships 193; and negotiated governance 186; services in private sector new towns 103
Local Development Frameworks 143, 165, 188, 189, 298
local nature reserves 146, 148
Local Strategic Partnerships 188, 190–1, 193
London 34; access to green belt 135; BedZED development 121–2; Docklands 42; Greater London Plan 1944 34, 36; green belt 126, 135, 165, 167; Molesey Heath 138; orbital road for 36, 81; outer ring area 30

London and Kent Thames Gateway 54, 111, 129–30, 199
London Wetlands Centre 153, 154
Lowe, M.S. 102, 103
Lowenthal, D. 76–7, 85
Lynch, K. 206

M25 36–7, 81
Maidenhead, Windsor and Eton flood alleviation scheme 111
Making of the English Landscape, The (Hoskins) 28–9, 77–8
management of the fringe (*see also* integrated management) 21, 177–8, 179
Manchester 55, 56–7, 96; City of Manchester Stadium 136; Trafford Centre 38, 40, 79, 97, 99
manufacturing industry 57; decline of 41, 42, 91
Matless, D. 28
Meadowhall (Sheffield) 40
mental hospitals 6, 73, 75
Metro Centre (Gateshead) 40, 100
Midlands 55, 67
military sites (*see also* airfields) 8, 59–62
mineral workings 13, 55, 64, 109–11, **110**; re-use of 65–6, 111–12
Molesey Heath 138
Morris, H. 168
Morton, N. 12
motorways 6, 29, **30**, 31, 32, 43, 73, **77**; as barrier to access 133; development associated with 29; hub housing development at junctions 96, 120, 124–5, 137; in North West England 96
multifunctionality 19–25, 93, 211, 212; barriers to 204; planning and 172, 182, 183; re-use of industrial sites 58, 60–1, 112; and sustainable development 155
Murdoch, J. 140
Myerson, G. 168

Nairn, I. 32–3, 78, 150
National Grid 151
Natural England 136, 193
nature 75, 77, 156–7; conservation of 142, 153; local reserves 146, 148
negative perceptions 82, 83, 88, 90
negative representations 17, 18, 80–1
neglect of fringe 4, 27, 28, 35, 45, 69
negotiated governance 183, 185–7

new economy: edge cities 102–5; and transport infrastructure 94–102
Nijkamp, P. 20
non-places 18, 44, 78–80
North West England 11, 83; road network 95–7
Northumbria Water 153
nuclear power stations 38, 86, **87**, 88

Office of the Deputy Prime Minister (ODPM) 170–1
open space 75, 116–17, 125–8, 139, 143, 144; educational benefits of access to 129–30; health benefits of access to 130–1
Orwell. George 29
Osment, F. 133, 156
out-of-town shopping (*see also* retail parks) 14–15, **41**, **79**, 93, 100, **101**; decentralization of retail 93, 99–100; political acceptability of 38–40

Pahl, R. 3–4
para-aesthetics 89–90
Parkhurst, G. 99
parklands *see* country parkland
parkway stations 14–15, 30, 97–8, 113
partnership working 23, 146, 181, 183, 185, 186, 202, 210; Community Forests 183–4, 193–4, 194–5; and integrated management 188, 190–2; local authorities and 193
paths 206
perceptions of the fringe 16–18; negative 82, 83, 88, 90
permeability of the fringe 206, 213
Pevsner, N. 88
Picon, A. 31–2
place-making 28, 62–3, 69
planning (*see also* land-use planning; spatial planning) 5, 13, 25–6, 35, 164, 212–13; archaeology and 53–4; and ecological fringe 139, 140, 143; and industrial heritage 58–9; post-war 162–3; reform of 187–96; role of in place-making 62–3; and sociocultural fringe 128, 129; uncoordinated 45
Planning and Compulsory Purchase Act 2004 187
planning delivery grants 209
'Planning for Real' workshops 187
Planning Policy Guidance Notes (PPGs) 53–4, 103, 170; access 131–2; archaeological heritage 54; development 103–4, 174; nature

conservation 145; out-of-town shopping 100; transport 105; waste management 64, 112–13
Planning Policy Statements (PPS) 140; on town centres 40; no.7 169–70, 171–2, 175, 177
play, opportunities for 129, 144
political fragmentation 21–2, 33–4
politics, and the fringe 38–40, 45, 208–9
pre-Roman settlements 52
preservationist groups 34
pressure to build 35–6
previously developed land 59
Prince, H. 76–7, 85
production at the fringe 91, 92
public-private partnerships 146, 147, 185, 209–10
public transport 92, 133, 206
Punter, J.V. 33, 150–1

quarries 64–5, **65,** 109, 111

railways (*see also* parkway stations) 6, 29, 30, 37, 97; marshalling yards 6, 73
Ravetz, J. 20–1
recreation (*see also* commercial leisure facilities; informal recreation): formal 14, 67, 68, 128
Red Rose Forest Trust 204–5
regeneration 21, 22–3, 43, 165; of derelict land 21–2; environmental 130, 147, 191, 204
Regional Spatial Strategies 190, 198
relocation of businesses 97, 99
representations of the fringe 16–19, 76–81, 207; negative 17, 18, 80–1
reservoirs 67, 126, 127, 154
residential development *see* housing
retail parks (*see also* out-of-town shopping) 4, 14–15, 73, 78–80, 93; architectural quality of 38; associated with residential development 115, 118
re-use: conservation through 55, 57–62, 63, 66; of historic buildings 50, 69
ribbon development 9, 15, 34, 74, 115
rights of way 132
roads (*see also* motorways) 14, 31, 94–5, 116, 131; access to 5, 6, 93; building 31, 36, 43, 45, 59, 94, 141; peri-urban 113, 118, 136; peripheral routes 206; ring roads 133; and urban growth 94

Roberts, B.K. 28
Rodenburg, C.A. 20
Rogers, Richard 119
romanticism 80
rough land 8, 74, 75
Rowe, P.G. 28
Royal Town Planning Institute 124, 203, 209; 'Modernising green belts' 173–6
rubbish tips 6, 73
rural buffers 92, 171, 177
rural landscape, impact of modernization on 77–8
rural-urban fringe 3–5; defining 5–11; future of 212 (visions for 184–5, 195, 202–3); making of 27–45; subdivision of 10
Rydin, Y. 168

safety, in the fringe 67, 68, 207
Samuel, R. 58
sanitization of the fringe 67, 128, 139, 183, 213
science parks 43, **44,** 93, 103, **104**
Section 106 Agreements 132–3, 153
sequential approach to site search 100–1
service functions 3, 5, 6, 73, 93, 113–14
service industry 42
sewage treatment works 4, 5, 6, 38, 73
Shirley, P. 141, 143
Shoard, M. 27–8, 41–2, 80, 84, 103, 131, 138, 139, 140, 143, 144, 145, 179, 200
shopping malls *see* retail parks
Sieverts, T. 89
Sinclair, I. 36, 81
Sites of Special Scientific Interest 145, 154
Smith, T.L. 31
sociocultural fringe 137, 179
sociocultural functionality 21, 22, 134, 183
Sontag, S. 88–9
South East England, housing development 59, 141, 164
South East Northumberland Forest Park 82–3
South East Plan 37, 198
space, need for 5, 6, 93
spatial planning 19, 182–7, 200; role of green belts 173–4
Statements of Community Involvement 187, 193
statutory undertakers 33, 150–1,

152–4; corporate social
responsibility 152–3
Stour Valley (Kent) 68
strategic gaps 165, 170, 171, 177
Stratford-upon-Avon 136, 207
subtopia 32–3, 78, 150–1
suburban development 34, 74, 120
Sullivan, W.C. 120
supermarkets 78, 101–2
sustainability 25, 53, 119, 212; public participation and 192
Sustainable Communities Plan 59
sustainable development 155; green belts and 173
sustainable living 122, 123–4
Sustrans 134
Swyngedouw, E. 85–6

Tate, R. 165
technoburbs 102
technophobia 17, 86
Tewdwr-Jones, M. 168–9, 182
Thames estuary *see* London and Kent Thames Gateway
Thames Water 154
Thayer, R. 17, 86, 88
Thomas, D. 3, 29, 52
tourist attractions 50
Town and Country Planning Act 1947 162, 196
Town and Country Planning Association 178; 'Policy Statement on Green Belts' 176–7
Trafford Centre (Manchester) 38, 40, 79, 97, 99
transitional landscape 4, 5, 9–10, 32–3, 74–5, 78
transport (*see also* car; railways): 94–102, 104–5; and economy of fringe 30, 43, 92–3
transport hubs 96, 100–1, 113, 141, 206; and housing development 96–7, 120
traveller sites 6, 73, 75

tree planting 149
Trinder, B. 28, 29, 37

United States of America 39, 102–3, 120
untidiness of land 35, 81, 82, 139
Unwin, R. 164
urban dowry 85–6, 88, 92
urban extensions 59, 74, 119, 141, 172
urban fringe 31
urban fringe action plans 197–8, 199
Urban Task Force 68, 119, 142, 163
utilities 150–1, 152–4

Van Diemen's Land, Bedford 83, 149, 150, 203
Vejre, H. 20
vernacular architecture 81–2
visions for the future 184–5, 195, 202–3

Wales 50, 55, 63, 145
warehousing 4, 14, 15, **17**, 61, 73, 93, 104
waste management 4, 38, 64, 109, 113–14; sites for 64, 65, 112–13
water-sports 6
Wellesbourne Airfield 24, 60, 205
wetlands 111, 153, **144**, 154
Whitehead, J.W. 12, 28, 165
Wibberley, G. 3
Wilce, H. 129
wildlife 75, 138, 145, 156; habitats 11, 111, 145, 156; protecting 139, 145, 153, 154–5
wind turbines 86, **87**
Wood, R. 20–1
woodland 111, 116, 127, 138, 149

York, archaeology of the fringe 52
Yorkshire 55, 56

zoning 20